ENERGY SCIENCE, ENGINEERING AND TECHNOLOGY

NATURAL GAS AND PETROLEUM

PRODUCTION STRATEGIES, ENVIRONMENTAL IMPLICATIONS AND FUTURE CHALLENGES

ENERGY SCIENCE, ENGINEERING AND TECHNOLOGY

Additional books in this series can be found on Nova's website under the Series tab.

Additional e-books in this series can be found on Nova's website under the e-book tab.

ENERGY SCIENCE, ENGINEERING AND TECHNOLOGY

NATURAL GAS AND PETROLEUM

PRODUCTION STRATEGIES, ENVIRONMENTAL IMPLICATIONS AND FUTURE CHALLENGES

CONSTANTIN CRANGANU
EDITOR

nova publishers
New York

Library of Congress Cataloging-in-Publication Data

Natural gas and petroleum : production strategies, environmental implications, and future challenges / editors, Constantin Cranganu (The City University of New York, Brooklyn College, NY, USA, and others).
pages cm
Includes bibliographical references and index.
ISBN 978-1-62808-535-8 (hardcover)
1. Natural gas. 2. Petroleum engineering. I. Cranganu, Constantin.
TN880.N2935 2013
665.5--dc23
2013026854

Published by Nova Science Publishers, Inc. † New York

CONTENTS

PREFACE

Today's oil and gas are at record prices, whilst global energy demand is increasing from population and economic development pressures. New and highly improved methods to detect, explore, and exploit new resources of oil and gas are necessary to maintain the world's energy needs toward sustainable economic development. This books aim is to offer the latest developments in several areas related to the above mentioned goals. As authors, it is our understanding and expectation, that interested readers will find plenty of useful, insightful, and reliable information for their energy-related tasks.

Chapter 1 – Hydraulic fracturing is the preferred stimulation method in shale oil and shale gas plays and is also used routinely to deduce *in situ* stresses. The principle behind the process is very simple, increase borehole fluid pressure until the rock fails by fracturing. Laboratory experiments have demonstrated that the actual process is much more complicated involving fluid diffusion and dependent on critical flaws. Hydraulic fracture experiments coupled with acoustic emission monitoring followed by SEM observations are providing new insights into the complexity of the process. SEM observations support the complexity, non-planarity and domination of shear failure interpreted from acoustic emission waves.

Chapter 2 – Integration, handling data of immense size and uncertainty, and dealing with risk management are among crucial issues in upstream oil and gas industry. The problems we have to solve in this domain are becoming too complex to rely on a single discipline for effective solutions, and the costs associated with poor predictions (e.g., dry holes) increase. Therefore, there is a need for establishing a new approach aiming at proper integration of disciplines (such as petroleum engineering, geology, geophysics, and geochemistry), data fusion, risk reduction, and uncertainty management in upstream oil and gas industry.

The authors discuss using several intelligent approaches (Artificial Neural Networks, Gene Expressing Programming, and Support Vector Regression) to provide a reliable methodology for estimating missing data (e.g., acoustic log, permeability, fluid pressures or fluid saturations), while maintaining procedures that are robust, converge fast, do not overpredict mean values and thereby preserve original data variability. Using the same data set as input variables, the authors will exemplify the use of three intelligent approaches in simulating missing sonic logs in the Anadarko Basin, Oklahoma and further use of those logs in estimating the presence, depth, and magnitude of abnormal pore-fluid pressures. The results are promising and encouraging.

Chapter 3 – Using mercury intrusion porosimetry technique the authors analyzed 91 core samples taken from the reservoir and non-reservoir rocks of the Anadarko Basin, Oklahoma.

The samples are sandstones (20), shales (24), interbedded shales and siltstones (11), siltstones (17), grainstones (9) and lime mudstones (10), ranging in age from Devonian to Permian and in depth from 404 m to 4,975 m. The objectives of this study are (1) to develop a classification of reservoir vs. non-reservoir rocks in the Anadarko Basin based on their lithology and petrophysical characteristics; (2) to establish the relationships between various petrophysical properties, lithology, and reservoir quality; 3) to determine how these relationships are affected by sorting of the pore-throat size distribution, and (4) to build a petrophysical database for reservoir and non-reservoir rocks of the Anadarko Basin. The results of this study will help resolve some of the key challenges faced in both exploration for and production of hydrocarbon: using only those petrophysical characteristics that can clearly distinguish between reservoir and non-reservoir rocks, identifying and characterizing petrophysical flow units, indicating the number of significant fluid invasions, understanding the relationships between petrophysical properties and lithology, or inferring the reservoir quality and performance.

The petrophysical parameters used in this study are: median pore-throat radius (MPR), average pore-throat radius (APR), apparent porosity (Φ), air permeability (Kair), pore-throat radius at 35% mercury saturation (R35), maximum threshold-entry radius (MTER), pore-throat size distribution (PTD), port type (PT), and pore-throat sorting (PTS). The quality and performance of reservoir rocks was estimated using a permeability/porosity/port type plot. Influence of lithology on petrophysical parameters was investigated. Finally, statistical analysis was used to assess correlations between various petrophysical parameters.

APR, MTER and Φ values are not well suited to discriminate between reservoir and non reservoir rocks (their value intervals mostly overlap). The authors assume that this situation is due to lithological heterogeneity of samples.

MPR, R35, and Kair have distinct value intervals for reservoir and non-reservoir rocks. The authors' interpretation is that the main control is exerted by port types: mostly mega and macro port types for reservoir rocks and mostly micro and nano port types for non-reservoir rocks.

Reservoir rocks display a majority (36 out of 50 samples) of bimodal pore-throat distributions, while all non-reservoir rocks (41 samples) have unimodal distribution. PT and PTS are considered responsible for this situation.

Another significant distinction between reservoir and non-reservoir rocks is provided by PTS: it is dominated by PS (poorly sorting) and MS (medium sorting) for reservoir rocks and by WS (well sorting) and MS for non-reservoir rocks. The link between PTS and lithology is clearly illustrated. Most samples that have WS PTS are sandstones (44%); most samples that exhibit MS PTS are siltstones (32%), whereas most samples with PS PTS are shales (33%). The distributions are significantly different ($\chi^2 = 23.606$, DF = 10, p-value = 0.009).

Two relationships (Φ vs. APR and Φ vs. MTER, both for reservoir rocks) have statistical significant correlation with p-value <0.05. None of statistical correlations has a statistical significance for non-reservoir rocks. The authors infer that lithological heterogeneity, depositional environments, lateral and vertical change in the formation's mineral composition and/or different port types may be responsible for such situations.

Although modest, the data collected during this research represent the beginning of a petrophysical database for reservoir and non-reservoir rocks of the Anadarko Basin, OK.

In this study the authors illustrated the various links between petrophysical characteristics and distribution of reservoir and non-reservoir rocks in the Anadarko Basin, Oklahoma.

This study produced (1) a classification of reservoir vs. non-reservoir rocks based upon their lithology and various petrophysical parameters; (2) identified criteria for classifying reservoir and non-reservoir rocks based on capillary-pressure curves; and (3) established statistical relationships between lithology and petrophysical characteristics of the reservoir and non-reservoir rocks in the Anadarko Basin, OK.

Chapter 4 – Natural gas (NG) and petroleum, the most important sources of energy around the world are discussed in this chapter. Relevant information regarding the NG and petroleum reserves, production and consumption and NG and oil prices is reported. The stock of proved reserves of natural gas distribution, in which Russia is the biggest holder, is included in this chapter. In case of petroleum, the evolution of proved reserves on each region is analyzed where most of the world oil reserves are in the Middle East. The distribution of proved reserves by region was analyzed to determine the ten top world reserves of Natural gas and petroleum and the prediction on what is expected in the coming years. As with other fossil fuels, burning natural gas produces carbon dioxide, which is the most important greenhouse gas. Since many scientists agree that high levels of greenhouse gas are changing the global climate, an analysis of the potential effect of increasing levels of carbon dioxide and other greenhouse gases in the earth's atmosphere is also presented.

Chapter 5 – Crude oil-in-water (O/W) emulsions have been proposed as an alternative and effective way to move viscous crude oils from oilfields to refinery and seaports, as well as across porous media inside the reservoir. O/W emulsion technology constitutes a lubricating flow method based on the dispersion of the crude oil as droplets within an aqueous phase containing emulsifying agents for reduction of the frictional energy loss produced during the viscous flow. The formation of a lubricating layer near to the pipe wall leads to a reduction of the frictional pressure drop and consequently to a decrease of the required energy for pumping. O/W emulsion flow represents also a productivity improvement on heavy oil pipeline by means of flow rate increase and the reduction of the maintenance and cleaning costs. This chapter discusses aspects of the formulation and characterization of crude oil-in-water emulsions for moving viscous crude oils. The main goal is to provide an overview of the main properties of petroleum disperse systems related to oil-in-water emulsions. It is also pointed out the effects of fluid dynamic, preparation and compositional variables on the stability, droplet size and viscosity of oil-in-water emulsion and their relevance for oilfield research and application of crude oil in water emulsions. Interfacial and colloidal aspects are highlighted and the role of natural surfactants from the oil on the emulsion properties is balanced. Technological outlook and practical experience in laboratory and oilfields were revised with support of the successfully worldwide research.

Chapter 6 – Petroleum and its derivatives have been considered important environmental contaminants. Among the petroleum derivatives, the monoaromatic hydrocarbons benzene, toluene, ethylbenzene and xylene (BTEX) are widely used as solvents and raw material in several industrial processes, moreover they are present in large amounts in fossil fuels. Due to their broad use and high solubility in water, BTEX is characterized as a frequent pollutant of waters, especially groundwater. Knowledge about the toxic effects associated to the exposure to the BTEX mixture is quite limited and this information refers mainly to the isolated effects of each one of these substances. In animals, these chemicals are easily absorbed, extensively metabolized and distributed to tissues highly vascularised and rich in lipids. Prolonged exposure to BTEX leads to renal, hepatic and lung toxicity, as well as alterations in the nervous and circulatory system. In certain concentrations, BTEX is genotoxic and mutagenic. The application of the microbial biodegradation process seems to constitute a promising methodology for the treatment of waters contaminated by this mixture. The use of bioindicators for the monitoring of environmental contamination has been considered quite satisfactory for identifying and assessing harmful effects of xenobiotics on test organisms, in their different concentrations and exposure periods, since these data can be extrapolated for organisms of the natural biota of impacted environments. Bioindicators also assist in the evaluation of the biodegradation process of environmental chemicals, allowing observing the effects caused by possible metabolites formed during this process. In this review, it will be presented studies related with the general characteristics of benzene, toluene, ethylbenzene, toluene and xylene and the biological action of these substances on exposed organisms. It will be also discussed the possibility of microbial biodegradation of BTEX and its application for remediation of groundwater contaminated by these xenobiotics.

Chapter 7 – This chapter presents a study of the cycle-to-cycle variations (CCV) in a natural gas spark ignition engine with and without exhaust gas recirculation (EGR). The presentation is divided into two parts. In the first part, no EGR is used, and the CCV of the indicated mean effective pressure (IMEP) are examined for different compression ratios (CR). The engine is operated at 1200 rpm and 1800 rpm, and four values of CR = 8, 10, 12, and 14 are used. In the second part, the EGR level is changed from 0% to 5%, 10%, 15% and 20%, at a fixed compression ratio of 9.4. The engine is operated at 2000 rpm and a stoichiometric fuel-air mixture is used. In both parts, the IMEP time series are analyzed using a continuous wavelet transform. The dominant oscillatory modes of the CCV are identified and the engine cycles over which these modes may persist are delineated. The results reveal that the CCV of the IMEP occur on multiple timescales and exhibit complex dynamics. With no EGR, high frequency intermittent fluctuations are observed at higher speeds. As EGR level is increased, persistent low frequency variations tend to develop. In addition, the spectral power increased with an increase in the EGR level. At the EGR level of 20%, the spectral power is found to increase significantly indicating that EGR has a pronounced effect on increasing the CCV. Knowledge of the dominant modes of variability may be useful to develop effective control strategies for reducing the CCV and improving engine performance in the presence and absence of EGR.

Chapter 8 – Over the past two decades, pore pressure prediction represents a key concept in hydrocarbon exploration and drilling. Encountering abnormal pore pressure while drilling is a common issue and has been reported by many drillers worldwide. Pressure kicks, borehole instability problems, stuck pipe, lost circulation, and even blowouts are typical occurring troubles while drilling formations with abnormal pore pressure. Therefore, a relatively reliable quantitative estimation of the pore pressure prior to drilling become imperative for a safe and economic drilling of new wells, especially in formations with potential overpressure risk.

The aim of this chapter is to provide reader with introductory topics in pore pressure prediction. It has been tried to accomplish this goal by avoiding complicated notations and advanced mathematics while, delivering accurate definitions for common petrophysical terms and methods used in pore-pressure prediction literature.

Historically, many geologists tried to classify and explore the geological mechanisms that may involve in producing high pore-pressure (e.g., Fertl, 1976; Fertl and Chapman, 1994).

Those efforts revealed that there are number of geologic processes which are commonly culprit in producing overpressure zones. There exist many other complex geologic systems where common proposed origins cannot fully address the true mechanism behind the observed pressure anomalies. However, number of novel models developed by frontier researchers to explain the observed anomalies within those reservoirs.

Generally, abnormal pore pressure is a product of geological, physical, chemical and mechanical processes.

Petrophysical and geological studies of various over pressured reservoirs from different part of the world confirmed that usually several pressure producing agents act together in order to generate overpressure within a reservoir.

Currently, geophysical methods especially reflection seismic techniques accompanied by non-seismic methods form a robust tool in pore-pressure estimation. Considering the fact that the fluid pressure has significant effect on compressional wave velocity, analyzing rock velocities became industry standard approach and its reliability confirmed by many geologists and drillers worldwide.

The following chapter contains four sections. Common petrophysical terms involved in pore-pressure studies will be introduced in the first section. Major geological overpressure origins will be discussed in section two. Section three covers the common accepted methods and relationships between different petrophysical parameters. Since rock velocities is the key parameter in many pore pressure estimation methods, understanding the nature of various velocity sources (e.g., velocities from seismic data both pre-stack and post-stack, VSP, logging, laboratory measurements, etc.) can improve our insight and reduce the associated pitfalls in using them. Accordingly, common velocity sources and their applications will be covered as a part of section three. Finally, in the last section of this chapter (section four) a full discussion on using velocity based methods in 3-D pore pressure estimation study provided as a practical example of using velocity based methods in pore pressure prediction studies.

In: Natural Gas and Petroleum
Editor: Constantin Cranganu

ISBN: 978-1-62808-535-8

Chapter 1

INSIGHTS FROM LABORATORY STUDIES OF HYDRAULIC FRACTURING

Carl H. Sondergeld, Chandra S. Rai, Yashwanth Chitrala* and Akash Damani
Mewbourne School of Petroleum and Geological Engineering
University of Oklahoma,
Norman, Oklahoma, US

ABSTRACT

Hydraulic fracturing is the preferred stimulation method in shale oil and shale gas plays and is also used routinely to deduce *in situ* stresses. The principle behind the process is very simple, increase borehole fluid pressure until the rock fails by fracturing. Laboratory experiments have demonstrated that the actual process is much more complicated involving fluid diffusion and dependent on critical flaws. Hydraulic fracture experiments coupled with acoustic emission monitoring followed by SEM observations are providing new insights into the complexity of the process. SEM observations support the complexity, non-planarity and domination of shear failure interpreted from acoustic emission waves.

INTRODUCTION

What is a hydraulic fracture? It is simply the use of a fluid to break bonds in a rock. The geometry of a borehole creates a magnification of applied far field stresses (Kirsch 1899) at the borehole wall. In deep formations, one principal stresses is taken to be vertical and two others are orthogonal horizontal stresses. Initial field experiments were carried out at shallow depths (Clark, 1949) leading to the false belief that all hydraulic fractures were horizontal. Hubbert and Willis (1957) recognized that hydraulic fracture orientations were controlled by in situ stresses and were predominately vertical with the fracture normal pointing in the

* now at BP America.

direction of minimum horizontal stress. Clearly then, the knowledge of in situ stresses becomes important in designing hydraulic fractures. Hydraulic fractures create new surface area and make reservoir drainage much more effective. Remote monitoring of this process is possible with limited resolution through pressure monitoring and/or through tilt meters and/or microseimics. With few exceptions (see Warpinski and Teufel, 1987; Jefferys et al., 2009) the stimulated fracture network is rarely sampled directly. At shallow depths, minebacks are feasible and reveal the complexity of the induced fracture network. Using microseismic event hypocenters, the fractures are mapped from surface arrays or observation wells. Similarly, tilt meters are used to measure displacements caused by the strain field associated with the hydraulic fractures. All these techniques suffer limitations of being indirect measurements which possess limited depths of investigation. Laboratory experiments provide control over many of the variables in a field scale experiment and permit direct observation of the actual hydraulic fractures. Scaling laboratory observations to the field depends on the physics of the processes being similar, i.e. the failure under laboratory conditions is that of the hydraulic fracture in the field.

Numerous papers have used the analysis of hydraulic fractures, minifracs and extended fracs (XLOT, DFITs) to deduce the magnitudes and directions of these stresses. In the simplest of cases, i.e. no leak-off into the formation, the breakdown stress, σ_{bd}, may be written as:

$$\sigma_{bd} = 3\sigma_h - \sigma_H + T_o - P_p \tag{1}$$

where σ_h is the minimum horizontal stress, σ_H is the maximum horizontal stress, T_o is the tensile strength of the rock and P_p is the pore pressure. If the two horizontal stresses are equal, the breakdown pressure becomes:

$$\sigma_{bd} = 2\sigma_h + T_o - P_p \tag{2}$$

This presumes that the failure mode is in tension. Laboratory studies of breakdown pressures in rock subject to hydrostatic loading do not follow eqn. 2. We have limited our focus to the development of tensile hydraulic fractures. The analysis leading to equations 1 and 2 yields nothing but stress magnitudes and directions and a measure of tensile strength. It reveals nothing about the hydraulic fracturing process, the size, density and type of fractures or the topology of the surfaces created. These details are reluctantly revealed in mineback experiments (Teufel and Warpinski, 1983; Jeffery et al., 2009) or field experiments designed to core through hydraulic fractures (Warpinksi et al., 1993; Fast et al., 1994; Moschovidis et al., 2000). Mineback experiments begin to capture the complexity of the actual fracture system and reveal in some examples how the fractures react to contrasts in elastic properties (Warpinski and Teufel, 1987). In several observations, the recovered cores capture remarkable thin induced fractures. Our understanding of this complex stimulation technique is quite rudimentary despite more than 60 years of implementation.

Many investigators have conducted laboratory studies of hydraulic fracturing (Hubbert and Willis, 1957; Haimson and Fairhurst, 1967, 1969; Rummel 1987; Lockner and Byerlee 1977,etc.). Early studies (Hubbert and Willis, 1957) established the relationship between fracture orientation and in situ stresses and addressed the breakdown pressures required for hydraulic fracturing. Haimson and Fairhurst (1967) studied breakdown pressures and fracture widths and proposed a formula for calculating fracture widths from known stresses, elastic properties and fracture propagation pressure. Rummel (1987) carried out experiments demonstrating the re-orientation of hydraulic fractures in response to changes in applied stresses. He also compared the magnitude of calculated and measured breakdown stresses, finding that measured breakdown stresses were much lower than predicted suggesting the existence of pre-existing cracks. Researchers have coupled acoustic emission monitoring with hydraulic fracture studies (Lockner and Byerlee, 1977; Sarda et al. 1988; Chitrala et al. 2010, 2011;Stanchits et al. 2011, 2012) and have detected pre-breakdown activity. Implicit in these studies is the assumption that observations and physics scale between the lab and the field. Both scales involve failure of rock which is dominated by shear failure (Warpinski et al. 2004; Chitrala et al. 2010, 2011; Damani et al. 2012). However, the process is modeled as a pure tensile Mode I failure (Yew, 1997). Field scale microseismic (MS) monitoring provides remote real-time mapping of fracture development and its geometry but suffers from remoteness, limited azimuthal coverage, and signal attenuation. The laboratory equivalent of MS monitoring, acoustic emission (AE) monitoring, records events in the frequency range of 50-1000 KHz compared to 200-2000Hz for MS (Middlebrook et al. 1993). This frequency difference implies a difference in the size of detectable acoustic sources; however, the failure process is fundamentally the same, i.e. the breaking of bonds. Rock failure processes in compression are better understood than in tension. Ultimate failure in compression is generally accepted to be result of culmination of a critical density of microfractures which link and form the final failure plane. The failure plane is defined by the highest density of microfractures and has a characteristic zone of decreasing fracture density orthogonal to the plane of the main fault. We infer, therefore, that the failure is surrounded by a "process zone." Of interest in hydraulic fracture stimulation are answers to the following questions: 1) Does such a process zone form during hydraulic fracturing; if so, under which conditions?, and 2) Is there a scaling relationship which defines the extent of the process zone? The answers to these questions are best obtained under conditions which can be controlled and where fractures can be observed.

Fracture mechanics models developed by Abou-Sayed et al. (1978) and Rummel and Hansen (1989) reproduce the episodic nature of hydraulic fracture propagation observed by Lockner and Byerlee (1977), Rummel (1987), Chitrala et al. (2010) and Damani et al. (2012). The balance between excess hydraulic energy and energy dissipated in creating new fracture surface area leads to the episodic propagation. The local stress concentrations at and around the crack tip cause fractures to deviate from a straight line path. SEM observations by Damani et al. (2012) and Bhagat (2012) (personnel communications) acoustic emission activity recorded by Lockner and Byerlee (1977), Chitrala et al. (2010) and Damani et al. (2012) support these fracture mechanics model. These finding suggest the surface roughness and hence the attendant fracture permeability would depend on the balance of available energy and newly created surface area.

Deductions of tensile strength from breakdown experiments find the predicted values to be less than independently measured values (Rummel 1987) or theoretically calculated values (Schmitt and Zoback, 1993). These observations led to the fracture mechanics models which couple the stress concentration of a circular borehole with those of a finite length crack subjected to various internal fluid pressure distributions. These models predict a decrease in tensile strength with increasing crack size (see Figure 1). They also predict a tensile strength dependence on the radius of the borehole. Thus, the differences between breakdown pressure and reopening pressure may be larger than predicted without independent knowledge of tensile strength. Further, initial defect size may be estimated directly (Rummel and Hansen, 1989).

These laboratory studies qualitatively confirm large scale field observations.

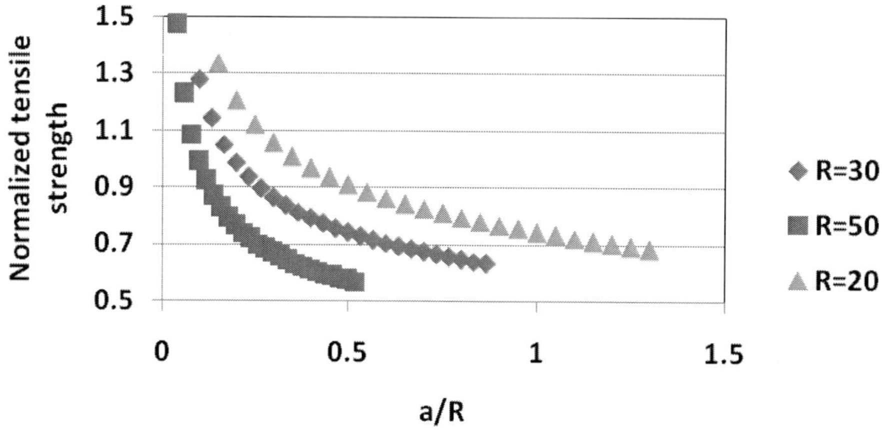

Figure 1. Normalized tensile strength plotted as a function of initial crack size, a, and borehole radius, R, expressed in mm. The expected tensile strength is taken to be the simple difference between breakdown pressure and fracture reopening pressure.

ACOUSTIC EMISSION STUDIES

Instrumenting experiments with piezoelectric transducers and recording the waves emanating from the hydraulic fracture permits nanoseismic or acoustic emission (AE) studies analogous to microseismic monitoring. Common to these techniques is the acoustic radiation of energy from a fracture. Frequencies and dimensions are different but material failure is common. With a known velocity model, we can triangulate and locate the source of the radiated energy (Lockner and Byerlee, 1977; Sondergeld and Estey, 1981) and through an analysis of the radiation pattern, we can infer with simple models, source dimensions (Aso, 2000) and mechanisms (Sondergeld and Estey, 1982). The temporal evolution of the spatial distribution of AE activity reveals the progression to failure, defining the zone of enhanced fracturing and the ultimate failure plane. Unfortunately, uncertainty in the locations precludes linking specific events to exact locations within the sample. Figures 3 A and B show the AE events located during a hydraulic fracture experiment of a uniaxially stressed sandstone. The sandstone has a porosity of 17% and a permeability of 20 μD. The fracturing fluid is oil with a viscosity of 50 cP and is injected at a depth of 50mm. A lateral stress of 1500 psi was applied in the direction of the arrows shown in Figure 3A. Sixteen piezoelectric sensors were used to capture the radiated energy. The bandwidth of the recording system and transducers was 50KHz to 2 MHz. Figure 2 shows the pressure build up in black (filled triangles) and the AE cumulative activity (green open triangles). The AE activity documents microfracturing activity prior to breakdown. Song et al. (2001) made a similar observation in experiments on

Tablerock sandstone. Including this information in the interpretation of stress magnitudes from hydraulic fracture provides critical input for realistic modeling of hydraulic fractures (Rummel and Hansen, 1989; Schmitt and Zoback, 1992) and can explain why observed tensile strengths in hollow cylinder tests are lower than those measured in other tests.

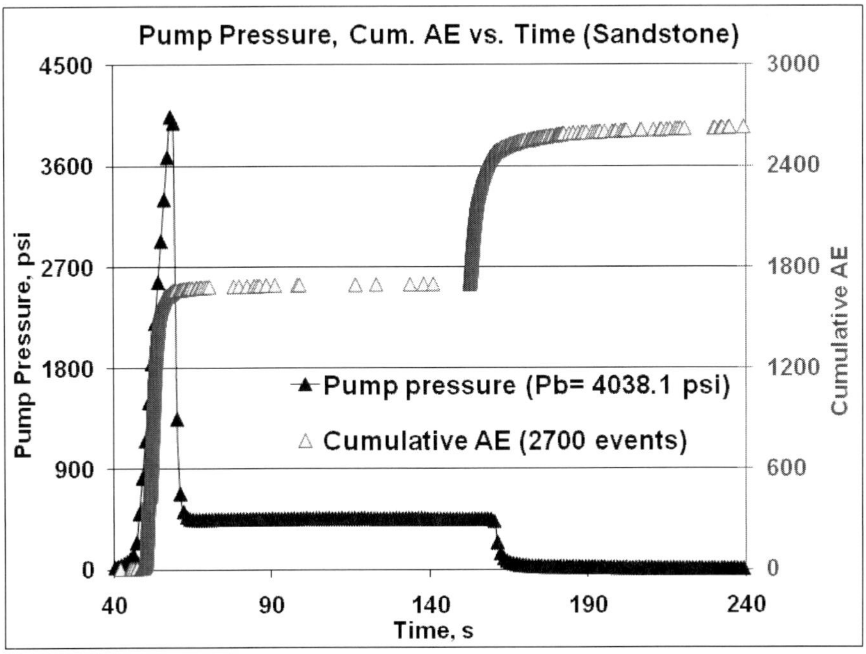

Figure 2. Pump pressure (black, solid triangles) and cumulative acoustic event activity (green, open triangles). Note there appears to be a considerable level of activity prior to breakdown and a marked increase in activity when the pump is shut down.

Figure 3A shows a plan view of a cylindrical sample (diameter = 104 mm) and located events colored by a time window in which they occurred. Figure 3B shows a lateral projection on to the X-plane. The only stresses applied are the borehole fluid pressure and the lateral stress, indicated by the red arrow. The borehole wall is not sealed so fluid leak off is possible but anticipated to minimal at the time of the failure. A number of features are obvious: 1) the activity aligns along a narrowly defined plane in the X-axis direction and in the direction of the maximum applied lateral stress; 2) the vertical projection onto the X-Z plane shows lateral growth over time; 3) latter events are more concentrated along the X-axis, and 4) fracture growth appears to be upward. Later events are associated with the fracture closure and are thought to be the result of asperity failure. We would therefore expect the locations to be concentrated along the existing fracture. A total of about 766 events were located from the 2600 recorded.

The cloud of events has an average width of about 7.5 mm. This is much larger than the grain size (~180 μm) which suggests there are strong deviations from a simple plane and/or uncertainty in hypocenter locations. Visual observation of the surface showed the actual fracture to be located in the middle of the AE cloud and to extend laterally beyond the AE locations along the X-direction. We can therefore map AE activity with limited resolution, in this case to within about ±4mm.

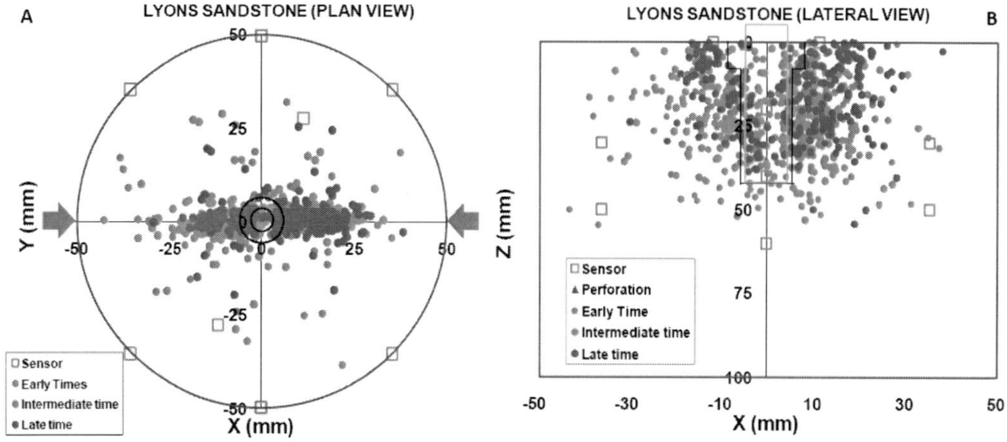

Figure 3. Plots showing plan and cross-sectional views of the hydraulically fractured Lyons sandstone specimen with the located AE events. Also shown in the cross-sectional view is the profile of the wellbore configuration. Sensor locations are indicated by the cyan squares and early, intermediate and late time events are indicated by green, orange and blue filled circles. The applied stress direction is indicated by the large red arrows. AE events indicate a well-defined plane develops which is oriented in the direction of the applied stress.

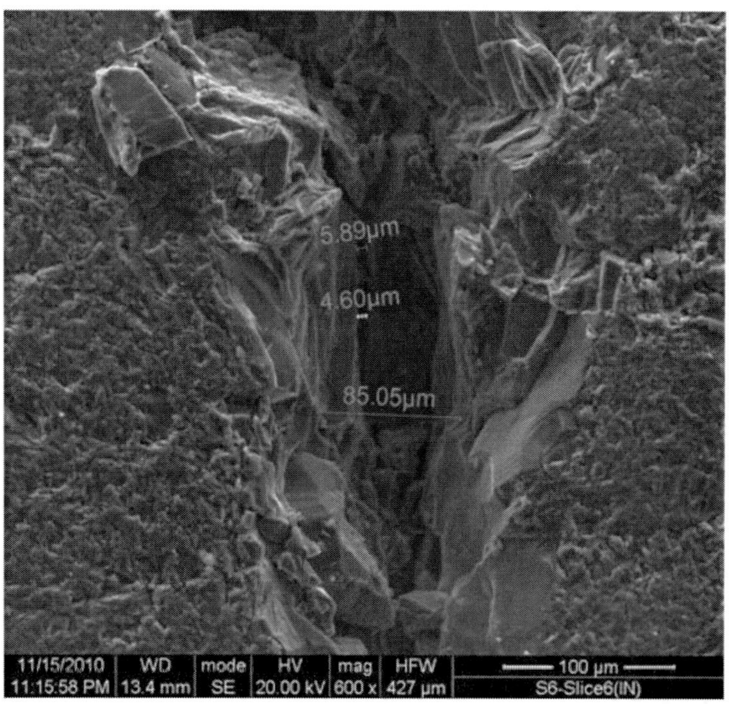

Figure 4. BSE image of Lyons sandstone hydraulic fracture. The near field shows a fracture width exceeding 100 µm; however, at the bottom of this imaged fracture we note dimensions of less than 5µm.

SEM images of the fractures show them to be narrow but complex (Chitrala et al. 2011). Figure 4 shows a back scattered electron image of a hydraulic fracture in Lyons sandstone annotated with dimensions. The near field portion of the image shows a relatively wide crack

(>85 μm) whose deeper features narrow to ~ 5μm. The wider portion of the fracture reflects the removal of loose and crushed grains. The narrow portion, we think, reflects the true stress relieved fracture aperture. Schmitt and Zoback (1993) remarked about the narrow cracks they observed in thin section analysis of hydraulically fractured hollow cylinders of Westerly granite. The actual fracture dimensions are 3 to 4 orders of magnitude smaller than the cloud mapped in Figure 3.

Figure 5 shows another location at the same depth in the sample at a lower magnification. What becomes obvious here is the complexity of the fracture, the shear offsets and fractured grains wedged in the main fracture. These fractures become self-propping which leads to enhanced permeability. Their tortuosity and surface roughness must impact proppant distribution. Fracture surface profiling reveals multiple topographic scales, most much larger than the measured grain size. Focal mechanism studies support the complexity observed in SEM images.

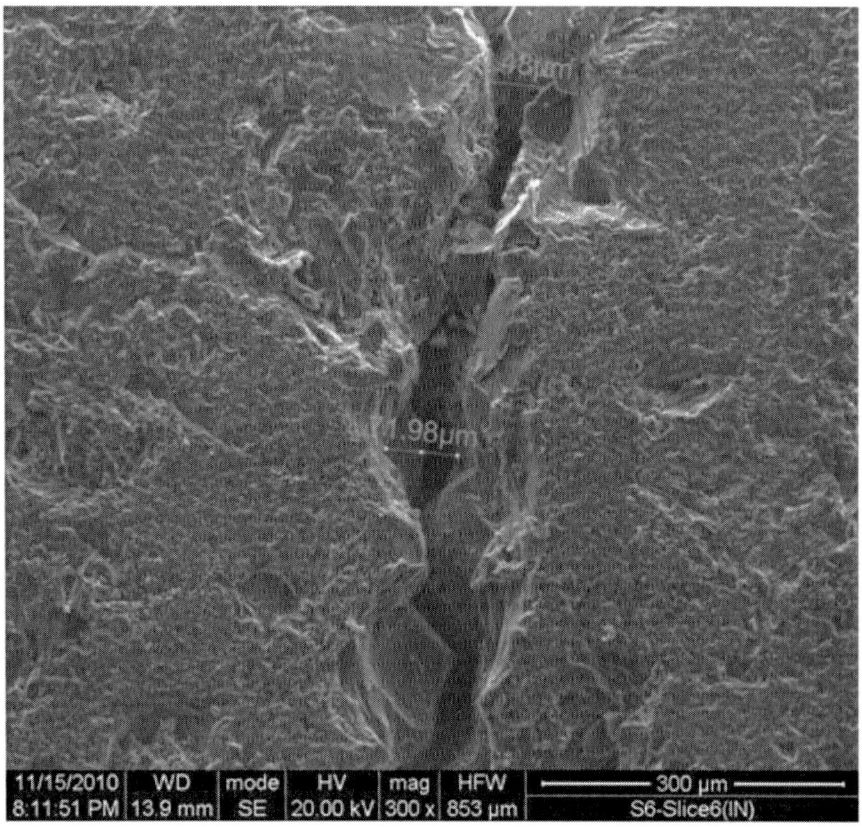

Figure 5. BSE image of Lyons sandstone hydraulic fracture. The magnification is less than that used in Figure 4. Fracture offset is clearly visible immediately below the 71.98 mm annotation. These offsets cause deviations from planarity, lead to local stress field changes and create tortuosity along the fracture surfaces.

Figures 6a and 6b show fracture surface roughness of a 5 mm by 5mm portion of the fracture as measured by an NANOVEA ST-400 optical profilometer. Figure 6A shows the raw topological data whereas Figure 6b shows only the higher frequency component. There are multiple scales of roughness created through hydraulic fracturing, the finest being limited by grain size in the slow propagation limit. Note that the color scales are different for the two images.

a b

Figure 6. Two dimensional optical profiles of a portion (5 mm x 5 mm) of the hydraulic fracture surface in a Lyons sandstone sample. Figure 6a shows the raw topography and 6b shows the same surface with the longer wavelength components removed. Note the color bars are different for the two images.

Fractal analysis of several fracture surfaces yields fractal dimensions which are on average 2.4 for five experiments conducted on Lyons sandstone. Resolution of the larger wavelength features is limited by the dimensions of the digitized surface. An example of the fractal analysis is shown in Figure 7; the fractal dimension is 2.56. The fractal dimension is an indication of the deviation from a pure planar feature which has a fractal dimension of 2.

Using the polarities of the P-wave arrivals and sensor locations relative to the hypocenters, we constructed lower hemisphere focal plots from which we deduced the nature of the source responsible for the observed polarities. We record with 16 sensors and require a minimum of 10 for focal mechanism determinations. The interpretation of the radiation pattern assumes the observations are made in the far field, i.e. 10 wavelengths from the source; we are minimally about 4 to 5 wavelengths away for some events. Nevertheless, we interpret polarities as if we are in the far field. Complex events are those which cannot be fit to a simple model; this is a consequence of noise and/or multiple overlapping events. Shear refers to those events which can be fit to a simple double couple model having clearly defined nodal planes.

Compression and tension are events for which a single positive or negative polarity is observed at all sensors. A summary of those analyses is presented in Figure 8 for two hydraulic fracture experiments in which the applied uniaxial horizontal stresses were 1000 psi and 125 psi, respectively. Statistically, shear is observed to dominate the recorded failure modes in both experiments. Shear events also dominate field observations (Warpinski et al., 2012). The complexity recorded in the focal mechanism analysis is supported by the SEM imaging of the fractures, e.g. see Figures 4 and 5.

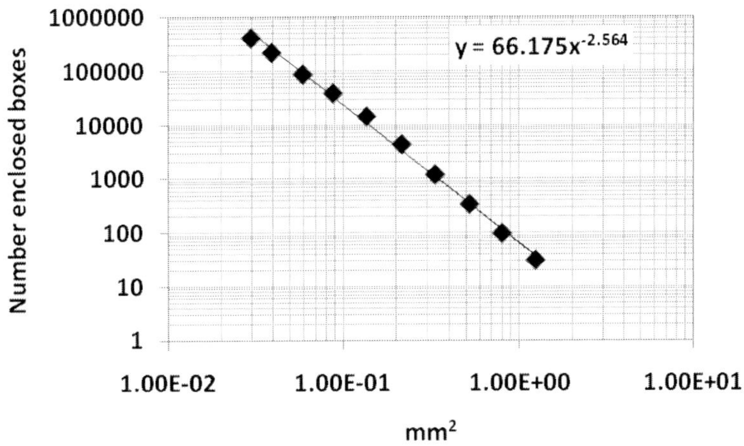

Figure 7. A fractal analysis of the surface presented in Figure 6. The fractal dimension is 2.56 which is considerably different than 2, a dimension associated with a simple planar surface.

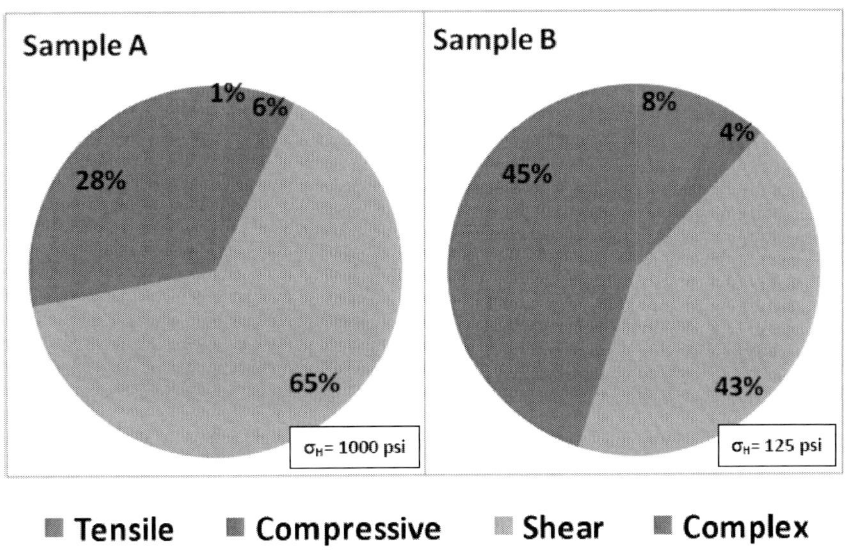

Figure 8. Focal mechanisms for AE events during uniaxial hydraulic fracture experiments. In either test with high or low applied stress, the dominant failure mechanism is observed to be shear.

CONCLUSION

Our studies indicate that lithology, anisotropy, pumping rate, propping fluid viscosity, deviatoric stress magnitude and orientation all influence the final stimulation. Brittle rocks produce more acoustic emissions than ductile rocks. Perhaps of prime importance is the confirmation that AE hypocenters accurately map the physically observed hydraulic fractures. The hypocenter cloud bounds the lateral extent, i.e. the width, of the fracture but not the length. Furthermore, focal mechanism studies and SEM observations confirm that many of the AE events result from shear failure. AE studies show clear premonitory activity before actual breakdown pressures are reached suggesting that the fracture actually initiates before breakdown. This may explain the discrepancy between measured tensile strengths and calculated tensile strengths using breakdown and reopening pressures and affects the interpretation of stress magnitudes from hydraulic fracture stress tests. Analysis of the polarities of the radiated sonic energy indicates that hydraulic fracturing is associated with shear failure. This is consistent with microseismic field studies. Fractures are tortuous and complex, not planar. Fracture propagation is episodic and not a monotonic continuous processes. Laboratory scale fracture morphology contains features observed on the field scale. Fracture speed varies from injection to termination which is reflected in the variation in fracture morphology along its path. At low deviatoric stresses, the rock fabric has more control over fracture network development and a more complex network is observed than at high deviatoric stresses. Increased confining pressure reduces the breakdown stress required for fracture initiation and simplifies the process zone. At higher magnitude of applied horizontal stresses, induced fractures are observed to align, as predicted, in the direction of maximum applied stress. Anisotropy influences the initial fracture direction and at low applied stress levels controls the hydraulic fracture direction. Unpropped fractures have permeabilities one thousand times greater than their virgin, unfractured permeabilities and display little pressure dependence.

The resultant fractures in laboratory experiments have roughnesses which are characterized with fractal dimensions of about 2.4, clearly different from simple smooth planes. Spatial frequency analysis detects fracture surface features many times greater than the intrinsic grain size. Observed fracture widths are narrow compared to their length and height with dimensions less than 100 μm and as fine as 5 μm. Through controlled experiments and post fracture analysis we will develop a body of knowledge which not only changes the interpretational framework for field microseismic studies but will allow more efficient hydraulic fracture designs and stress mapping.

REFERENCES

Abou-Sayed, A. S., Brechtel, C. E., Clifton, R.J., (1978). In situ stress determination by hydrofracturing: A fracture mechanics approach, *Jour. Geophys. Res.* 83, B6, 2851-2862.

Aso, I. I., (2010). Microseismic mapping of hydraulic fractures, *MS Thesis,* The University of Oklahoma, Norman, Oklahoma.

Chitrala, Y., Moreno, C., Sondergeld, C., Rai, C., (2010). *Microseismic mapping of laboratory induced hydraulic fractures in anisotropic reservoirs*, SPE-138441, San Antonio, Texas, Nov 2-3.

Chitrala, Y., Moreno, C., Sondergeld, C., Rai, C., (2011). *Microseismic and microscopic analysis of laboratory induced hydraulic fractures,* SPE-147321.

Clark, J. B., (1949). A hydraulic process for increasing the productivity of wells, *Trans. AIME*, Vol. 186, 1-8, SPE949001.

Damani, A., Sharma, A., Sondergeld, C., Rai, C., (2012). Mapping of hydraulic fractures under triaxial stress conditions in laboratory experiments using acoustic emissions, SPE159604, *presented at the annual SPE ATCE*, San Antonio, 8-10 Oct.

Fast, R. E., Murer, A. S., Timmer, R.S., (1994). Description and core analysis of cored hydraulic fractures, Lost Hills Field, Kern County, California, *SPE Production and Facilities*, May, pp. 107-114.

Haimson, B., Fairhurst, C., (1967). Initiation and extension of hydraulic fractures in rocks, *Soc. Pet. Eng. J.* 7, pp. 310-318.

Haimson, B., Fairhurst, C., (1969). Hydraulic fracturing in porous-permeable materials, *J. Petrol. Tech.* 7, pp. 811-817.

Hubbert, M. K., Willis, D. G., (1957). Mechanics of hydraulic fracturing, Petroleum Transactions, *AIME*, 210, pp. 153-166.

Jeffrey, R. G., Bunger, A. P., Lecampion,B., Zhang, X., Chen, Z. R., A. van AS, Allison, D. P., de Beer, W., Dudley, J., Siebrits, E., Thiercelin, M., Mainguy, M., (2009). *Measuring hydraulic fracture growth in naturally fracture rock*, SPE 124919.

Kirsch, G., (1898). Die Theorie der Elastizität und die Bedürfnisse der Festigkeitslehre, *Zeit. Ver. dt.Ingenieure*, 42, 797-807, 1898.

Lockner, D. A., Byerlee, J. D., (1977). Hydrofracture in Weber sandstone at high confining pressure and differential stress, *J. Geophys. Res.*, 82, 14, pp. 2018-2026.

Lockner, D. A., Byerlee, J. D., Kuksenko, V., Ponomarev, A., Sidorin, A., (1992). *Observations of quasistatic fault growth from acoustic emission, in Fault Mechanics and Transport Properties of Rocks,* ed. by B. Evans and T-F Wong, Academic Press, Sa Diego, 523 pp.

Middlebrook, M., Peterson, R., Warpinski, N., Engler, B., Sleefe, G., Cleary, M., Wright, T., Branagan, P., (1993). Multi-site project seismic verification experiment and assessment of site suitability, *GRI-report*,93/0050.

Moschovidis, Z., Steiger, R., Peterson, R., Warpinski, N., Wright, C., Chesney, E., Hagan, J., Abou-Sayed, A., Keck, R., Frankl, M., Fleming, C., Wolhart, S., McDanel, B., Sinor, A., Ottesen, S., Miller, L., Beecher, R., Dudley, J., Zinno, D., Akmedov, O., (2000). *The Mounds drill-cuttings injection field experiment: Final results and conclusions,* SPE 59115.

Rummel, F., 1987, Fracture mechanics approach to hydraulic fracturing stress measurements, in Fracture Mechanics of Rock (Edited by B.K. Atkinson), Academic Press, London, pp. 217-239.

Rummel, F. Hansen, J., (1989). Interpretation of hydrofrac pressure recordings using a simple fracture mechanics simulation model, *Int. J. Rock Mech. Min. Sci. &Geomech.Abstr.* 26, 6, 483-488.

Sarda, J-P., Perreau, P. J., Deflandre, J-P., (1988). *Acoustic emission interpretation for estimating hydraulic fracture extent: Laboratory and field studies*, SPE 13192.

Schmitt, D. R., Zoback, M. D., (1992). Diminished pore pressure in low-porosity crystalline rock under tensile failure: Apparent strengthening by dilatancy, *J. Geophys. Res.* 97, 1, 273-288.

Schmitt, D. R., Zoback, M. D., (1993). Infiltration effects in the tensile rupture of thin walled cylinders of glass and granite: Implications for the hydraulic fracturing breakdown equation, *Int. J. Rock Mech. Min. Sci. & Geomech. Abstr.*30, 289-303.

Sondergeld, C. H., Estey, L. H.,(1981). Acoustic emission study of microfracturing during the cyclic loading of Westerly granite, *J. Geophys. Res.,* 86, B4, 2915-2924.

Sondergeld C. H., Estey, L. H., (1982). Source mechanisms and microfracturing during uniaxial cycling of rock, *PAGEOPH*, 120, 151-166.

INTRODUCTION

Integration, handling data of immense size and uncertainty, and dealing with risk management are among crucial issues in upstream oil and gas industry. The problems we have to solve in this domain are becoming too complex to rely on a single discipline for effective solutions, and the costs associated with poor predictions (e.g., dry holes) increase. Therefore, there is a need for establishing a new approach aiming at proper integration of disciplines (such as petroleum engineering, geology, geophysics, and geochemistry), data fusion, risk reduction, and uncertainty management in upstream oil and gas industry. In this chapter we wish to explore a methodology, called intelligent approaches in upstream oil and gas industry, that might address most of the concerns expressed above and offer a new form of integrating the disciplines rather than integrating the results.

Intelligent approaches such as neural computing, fuzzy reasoning, and evolutionary computing for data analysis and interpretation are an increasingly powerful tool for making breakthroughs in the science and engineering fields by transforming the data into information and information into knowledge. In the oil and gas industry, these intelligent techniques can be used for uncertainty analysis, risk assessment, data fusion and mining, data analysis and interpretation, and knowledge discovery, from diverse data such as 3-D seismic, geological data, well logging, and production data (Nikravesh and Aminzadeh, 2003).

Intelligent approaches may be defined as a consortium of computing methodologies that collectively attempt to imitate life and has matured to a set of analytic tools that facilitate solving problems which were previously difficult or impossible to solve. The role model for intelligent approaches is the human mind.

Some of the prominent members of the consortium are: Artificial Neural Networks (ANN), Genetic Algorithms (GA), and Support Vector Machines (SVM). They may be grouped together to create powerful Committee Machines (CM). The main goal of this chapter is to provide a reliable methodology for estimating missing data (e.g., acoustic log, permeability, fluid pressures or fluid saturations), while maintaining procedures that are robust, converge fast, do not overpredict mean values and thereby preserve original data variability. The proposed methodology also deals greatly with uncertainty associated with the data, the immense size of the data, the diversity of the data type, and will be a valuable approach to future exploration and exploitation studies in oil and gas industry.

Eventually, our methodology can be extended to other problems encountered in upstream oil and gas industry and can be a key to cost effectively locating and producing our remaining hydrocarbon reserves.

The remaining chapter will present in detail three intelligent approaches drawing upon on our previously published works: Artificial Neural Networks (ANN) (Cranganu, 2007); Gene Expressing Programming (GEP) (Cranganu and Bautu, 2010), and Support Vector Regression (SVR) (Cranganu and Breaban, 2013). For the first time, these three methods will present as case studies the same problem: estimating missing sonic logs (DT) from well logging suites recorded in the Anadarko Basin, OK, in order to estimate the presence, location, and magnitude of abnormal pore-fluid zones (overpressures). Although limited by typographic space to present only one example of using intelligent approaches in upstream oil and gas industry, it is easy to expand these approaches to many other problems (see above for an incomplete list).

METHODS

1. Artificial Neural Networks (ANN)

Many sedimentary basins throughout the world exhibit areas with abnormal pore-fluid pressures (higher or lower than normal or hydrostatic pressure). Predicting pore pressure and other parameters (depth, extension, magnitude, etc.) in such areas are challenging tasks. The compressional acoustic (sonic) log (DT[1]) is often used as a predictor because it responds to changes in porosity or compaction produced by abnormal pore-fluid pressures. Unfortunately, the sonic log is not commonly recorded in most oil and/or gas wells.

We propose using an artificial neural network to synthesize sonic logs by identifying the mathematical dependency between DT and the commonly available logs, such as normalized gamma ray (GR[2]) and deep resistivity logs (REID[3]). The artificial neural network process can be divided into three steps: (1) supervised training of the neural network; (2) confirmation and validation of the model by blind-testing the results in wells that contain both the predictor (GR, REID) and the target values (DT) used in the supervised training; and 3) applying the predictive model to all wells containing the required predictor data and verifying the accuracy of the syn-thetic DT data by comparing the back-predicted synthetic predictor curves (GRNN, REIDNN) to the recorded predictor curves used in training (GR, REID).

Artificial neural networks offer significant advantages over traditional deterministic methods. They do not require a precise mathematical model equation that describes the dependency between the predictor values and the target values and, unlike linear regression techniques, neural network methods do not overpredict mean values and thereby preserve original data variability. One of their most important advantages is that their predictions can be validated and confirmed through back-prediction of the input data.

This procedure will be applied to predict the presence of overpressured zones in the Anadarko Basin, Oklahoma.

1.1. Introduction to ANN

It is well documented (e.g., Huffman and Bowers, 2002; Mitchell and Grauls, 1998; Surdam, 1997; Ortoleva, 1994) that in many sedimentary basins the pore fluid pressure measured at a certain depth is either higher (overpressure) or lower (underpressure) than the normal (hydrostatic) fluid pressure at that depth. A sedimentary basin experiencing one or both of these situations is said to have abnormal pressure regimes. Overpressured zones are far more common than underpressured zones.

Abnormal pore fluid pressure and abnormal pressure regimes have always been recognized as important factors in the evolution of hydrocarbon provinces (e.g., Huffman and Bowers, 2002; Poston and Berg, 2002; Mitchell and Grauls, 1998). Until recently, however, the ability to predict their presence, location, magnitude and distribution has been limited by two critical factors: our limited knowledge of those regimes and their physics, and the lack of sufficiently robust technologies for predicting pressures in the subsurface. Historically, abnormal pressure evaluation has used basic data such as mud weight, repeat formation

---

[1] Propagation time of seismic waves in and around a borehole, in μs/ft.
[2] Natural gamma radiation, in uAPI.
[3] Electric resistivity variation, in Ωm.

testing (RFT), and drill-stem testing (DST). Seismic velocities, wireline logs, and resultant porosity estimations can provide an indirect method of evaluating pore pressure.

The compressional acoustic (sonic) log (DT) has been used to characterize areas with abnormal pore fluid pressure via effective rock stress estimations. The DT log seems to be preferred due to its accuracy and sensitivity to changes in rock stress produced by abnormal pore-fluid pressures, especially in shale layers (Asquith and Krygowski, 2004; Hearst et al., 2000). Determining trends in acoustic transit time on sonic logs has become a conventional method of detecting anomalous pressures in shales. Because the velocity of a compressional wave in sed-imentary rock is dependent on the effective rock stress, an increase in transit time is interpreted as indicating a lower effective rock stress and, therefore, overpressure (Surdam et al., 1997).

The calculation of abnormal pressure from log data is based on the relationship between observed acoustic values of shales and a trend line reflecting normal compaction (see e.g., Heppard et al., 1998; Hsu and Balaud, 1998).

In spite of its intrinsic value for studying and analyzing abnormal fluid pressures, the sonic log is not commonly available in oil and gas wells. This may happen either because a full suite of logs is not recorded or because of problems encountered in repeat logging, such as damaged, faulty logging instruments, or poor logging conditions in any of the logging runs. As a result, some wells may lack the sonic log entirely or the log is only partially recorded.

Artificial neural networks are used to accurately synthesize missing DT data, increase the density of the control data, and more effectively map changes in pore pressure in the Anadarko Basin, Oklahoma, a basin well known for its overpressured compartments (Figure 1) (Al-Shaieb at al., 1990, 1997a, 1997b, Cranganu, 2004).

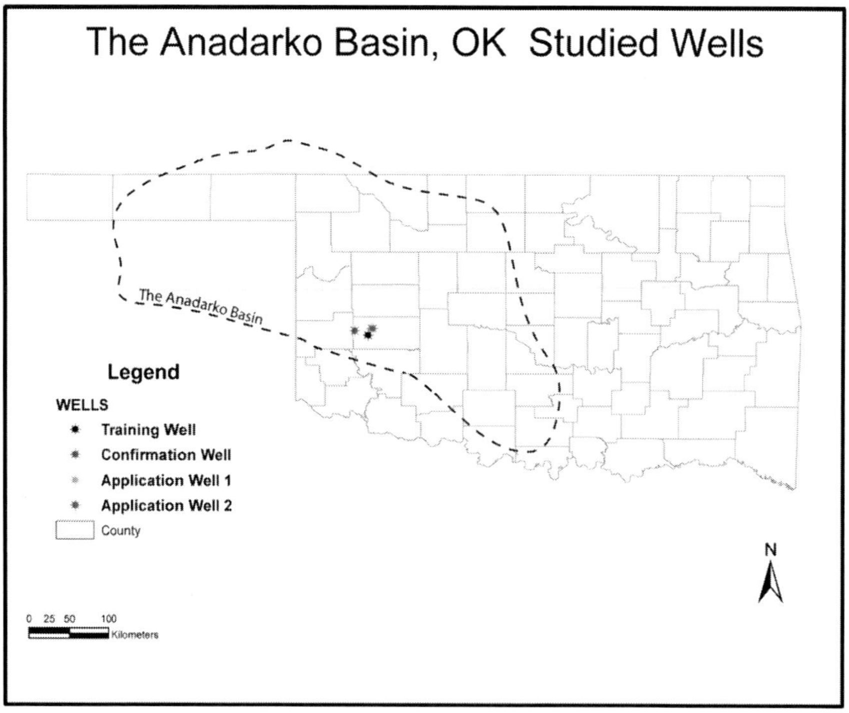

Figure 1. The Anadarko Basin, Oklahoma, and the wells used in the artificial neural network.

1.2. Artificial Neural Networks – Background

Artificial neural networks are computational systems that mimic the biological neural networks of the mammalian brain. The human brain contains about 100 billion neurons (neural cells), interconnected in a complex manner via synapses (a junction between axons and den-drites, thus constituting a network. Artificial neural networks are grossly simplified models of living neural networks (Rogers et al., 1995). They are also an exploration and development tool that allows an easy transformation of log data into any desired output parameter. Specific mathematical relationships between predictors (logs) and target values (rocks) need not be known.

Since the artificial neural networks mimic the human brain's problem-solving processes they can use knowledge gained from past experiences and apply that knowledge to new problems and situations. The artificial neural networks, in effect, use a training experience to build a system of neurons and weight links that allow them to make new decisions, classifications, and predictions (Arbogast et al., 2000).

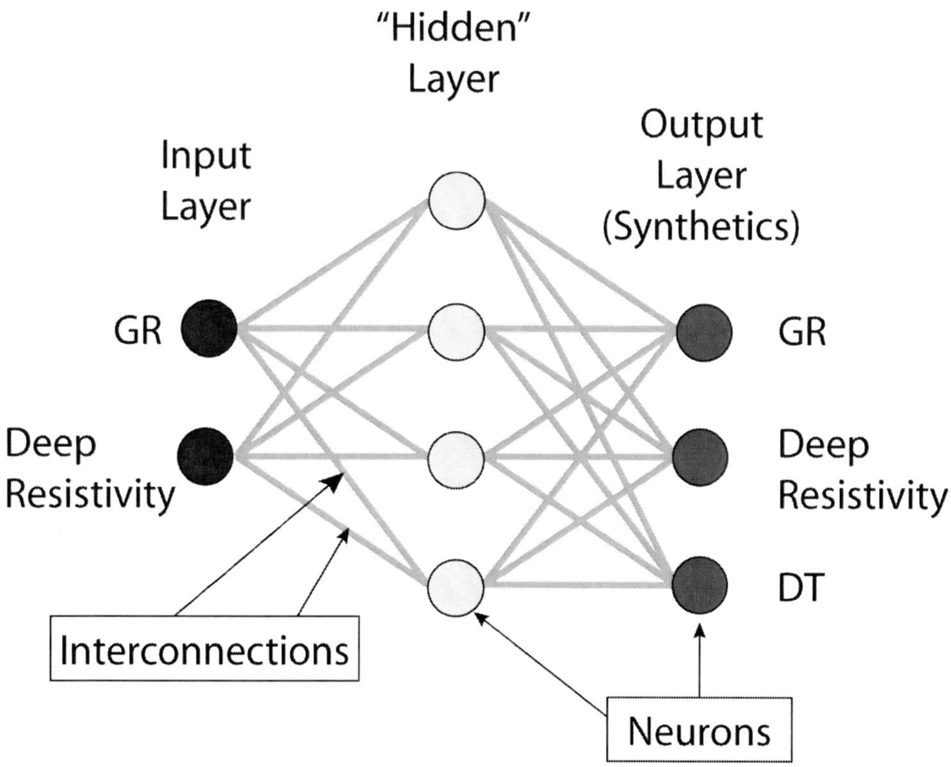

Figure 2. Artificial neural network architecture used in the Anadarko Basin, Oklahoma (from Cranganu, 2005a and Cranganu, 2006).

To be able to mimic the human's brain neural system, it is necessary for an artificial neural network to comprise (Figure 2) (*i*) a set of nodes (artificial neurons) where the nodes perform simple computations, (*ii*) a set of interconnections or synapses linking pair of nodes, and (*iii*) a set of labels known as *weights*, associated with each interconnection and identifying some property of interconnection. These weights correspond to the synaptic efficiency of the biolog-ical neurons (Aristodemou et al., 2005).

The neurons in the network perform two simple tasks. The first is to build a weighted summation of the input data (Figure 3). Specifically, a signal x_j at the input of the synapse j connected to neuron k is multiplied by the synaptic weight w_{kj}. A summing junction, Σ, acts as an adder for summing the input signals, weighted by the respective synapses of the neuron.

The second task is to apply a function to this summation to yield an output than can serve as an input to other neurons or that can go to the output layer (Figure 3). An *activation function* limits the amplitude of a neuron in that it squashes the permissible amplitude range of the out-put signal to some finite value. The neuronal model of Figure 3 also includes an externally applied *bias (threshold)* b_k that increases or decreases the net input of the activation function, de-pending on whether it is positive or negative, respectively.

Following Haykin (1999), we may describe a neuron k by writing the following pair of equations:

$$u_k = \sum_{j=1}^{n} w_{kj} x_j \tag{1}$$

and

$$y_k = \phi(u_k + b_k) \tag{2}$$

where $x_1, x_2, ..., x_n$ are the input signals; $w_{k1}, w_{k2}, ...w_{kn}$ are the synaptic weights of neuron k; u_k is the *linear combiner output* due to the input signals; b_k is the bias; $\phi(\bullet)$ is the activation function; and y_k is the output signal neuron. The weights w_{kn} associated with each connection indicate the extent to which the conveyed signal is amplified or diminished. The activation function $\phi(\bullet)$ is often a sigmoidal (S shaped) function, but other functions have been used (Heaviside or threshold or piece-wise linear). Given a network whose weights are initially random and assuming that we know the task needed to be accomplished by the network, a *learning algorithm* is required in order to determine the values of the weights that will achieve the desired task.

Considering an artificial neural network, three types of neurons can be distinguished de-pending on their function (Figure 2): (*i*) the input neurons, found in input layer, where the input value are given; these units do not have input connections but only output ones; (*ii*) the "hidden" neurons, which have both input and output connections; they vary in number and could be arranged in one or more "hidden" layers; and (*iii*) the output neurons, which have only input connections, and are grouped in the output layer.

A signal is input into the input layer from an external source and is propagated to the next layer by the interconnections between the neurons. The signal is processed and then sent to the next layer. The processing and then transmitting of the signal continues until the output layer is reached. There can be several layers of "hidden" neurons between the input layer and the output layer. This is an example of a *supervised multilayer feed forward* (MFF) with *back propagation* (BP) of error algorithm neural network.

Among the various types of artificial neural networks used today, it seems that the most widely accepted are those of MFF-BP type. Their application has been documented in numerous published papers dealing with estimating lithology (Rogers et al., 1995; Benaouda et al., 1999; Saggaf and Nebrija, 2000), porosity and permeability (Rogers et al., 1995; Huang et al., 1996; Schuelke et al., 1997; Wong et al., 2000, Ali and Chawathe, 2000, Ligtenberg and Wan-sink, 2003), fractured reservoir characterization (Zellou and Ouenes, 2003), and other reservoir rock properties (Baldwin et al., 1990; Accarain and Desbrandes, 1993; Saggaf and Nebrija, 2003).

In our work, we used a MFF-BP type of artificial neural network as it was implemented in a software called NNLAP (Neural Network Log Analysis Program) (Arbogast et al., 2000)

1.3. Artificial Neural Network – A Three-Step Technology

The operation of the artificial neural network used to determine the presence of overpressured regimes in the Anadarko Basin, Oklahoma, can be divided into three steps:

Step 1. Supervised training of the neural network by using a training well

The artificial neural network begins the training process by randomly assigning the initial interconnection weights between the layers of neurons. Then an input signal is given to the first layer of processing elements (the input layer, Figure 2), whereas a corresponding output re-sponse is presented to the last layer of processing elements. All connections between the input layer, the "hidden" layers, and the output layer are then adjusted using an objective function (or "cost function") such as the *mean squared error* (*MSE*) (Arbogast et al., 2000), the *sum squared error* (*SME*) (Aristodemou et al., 2005) or the *global error E* (Luhti and Bryant, 1997). For example, the *global error E* is computed as the sum of the squared differences between the computed output y_k and the desired output d_k (i.e., the local errors). For iteration n, the global error $E(n)$ is thus:

$$E(n) = \frac{1}{2}\sum_{k=1}^{n}\left\| d_k(n) - y_k(n) \right\|^2 \qquad (3)$$

In back propagation (BP) networks, the weight adjustment for every connection is computed in a gradient descent method in weight space, whereby the local gradient is calculated and a correction to the previous weight is made through multiplication of $\partial E / \partial w_{jk}$ with a *learning rate n* and a momentum α according to the formula (modified from Arbogast et al., 2000)

$$w_{jk}(t) = -n\frac{\partial E}{\partial w_{jk}} + \alpha \cdot \Delta w_{jk}(t-1) \qquad (4)$$

where t is the time step and Δw is the weight change from one cycle to another.

The learning rate n is a small number ($.1<n<1.0$) (Aristodemou et al., 2005) that controls the amount of the error that will be added, negatively, to the interconnection weights for the next cycle. If the learning rate is large, then large charges are allowed in weight changes. Conversely, if the learning rate is small, only small changes are allowed, which can increase learning time.

Momentum α is a term that dampens the amount of weight change Δw by adding in a portion of the weight change from the previous cycle. The addition of momentum is credited with smoothing out wild changes of weights, and also with helping the network converge faster when error is successively changing in the correct direction. Typical values for momentum fall between 0 and 1.0 (Aristodemou et al., 2005).

The local error on every connection is used to calculate corrected activity levels in the processing elements of the previous layer, which in turn are used to calculate new local errors. Thus, the error gets propagated backward through the entire network and new connection weights are obtained with every step. For a good derivation of the back propagation algorithm consult, for example, Haykin (1999).

In our training work with NNLAP program the *training settings* were chosen as follows:

- learning rate = 0.5
- momentum = 0.9
- stopping error = 0.0001
- training iterations = 400

The main *design settings* included the number of hidden layers (2 or 4, Figure 4) and the number of neurons per "hidden" layer (16). The training and design settings mentioned above were determined from trial-and-error experimentation during which an optimum network performance was sought. Attempts to varying these values did not produce discernable differences in output results. We found that more important in obtaining robust and consistent results was selection of the input curves to be used in training and also the location of the representative training examples. The NNLAP program stops the learning process when the local errors have dropped to a satisfactory level or when a reasonable number of training iterations has been reached. Limiting the number of training iterations can prevent "over-training."

NNLAP program has two types of training:

(1) *Quick Training* starts with a new neural network (new random numbers on all of the in-terconnection weights, etc.) and then it trains the neural network until, at the training examples chosen, the error is 0.0001 or 400 iterations have been made (whichever comes first). We found that having the possibility of choosing your own training examples is a great asset of the program and may reduce substantially the training time.

(2) *Continued Training* starts where the last training was left off and trains for an addition-al 400 iterations or until the error is 0.0001 (whichever comes first). If *Quick Training* provides a fairly good solution, it is unlikely that *Continued Training* will make major improvements. It might, however, "over-train" the neural network and start learning noise and other non-desirable parameters. Less is almost always better than more when using neural networks. The fewer input curves and the fewer training examples required to get a good solution in the training well, the more robust the solution will be when used in the confirmation and application wells.

Figure 5 shows the training well T & T 1-10 between 4,000 and 14,000 ft (1,219 and 4,267 m, respectively) and the curves used for training: GR (gamma-ray), REID (deep resistivity), and DT (sonic) logs. The training curves are shown in blue or yellow color. The calculated (synthetic) curves have a NN or N1suffix (GRNN, GRN1, REIDNN, REIDN1, DTNN, and DTN1) and displayed in red color. The six green horizontal lines represent the training examples. The black curve CAL[4] (caliper) was not used as a predictor curve but only as a reference curve so that training could be avoided in zones with invalid data caused by washouts, mud cakes, etc.

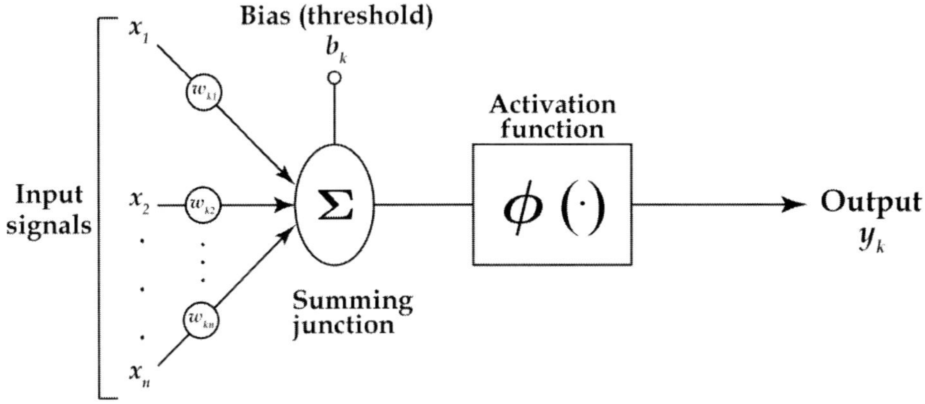

Figure 3. Non-linear model of a neuron (modified from Haykin, 1999).

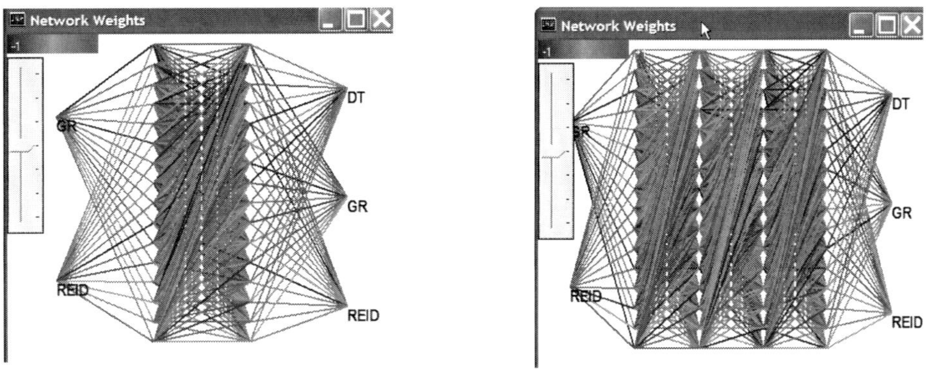

Figure 4. The structure of the artificial neural network used in the Anadarko Basin, Oklahoma: (a) with two “hidden” layers and 16 neuron in each layer; (b) with four “hidden” layers and 16 neuron in each layer.

We choose GR and REID as input training curves to simulate DT curves because these were the most common logs in the data set available to us. Additionally, we used Data Mining menu of NNLAP software to decide over our selection of input curves. In Data Mining menu, the user defines the output (target) curve and puts a check mark in the boxes next to all of available input curves and then the program will rank the available curves based on their corre-lation coefficient relative to the target curve. After running the program, we found that an accurate selection of a particular log as the network input curve was not necessary because using multiple logs seems to give equivalent results as when using the highest correlated log. Similarly, Saggaf and Nebrija (2003) found that using various input

[4] Measures the borehole diameter variations, in inches.

log sets to simulate DT (e.g., GR+ Neutron Porosity, GR + Density, Neutron Porosity + Density) did not yield significantly different results in the simulated DT curves.

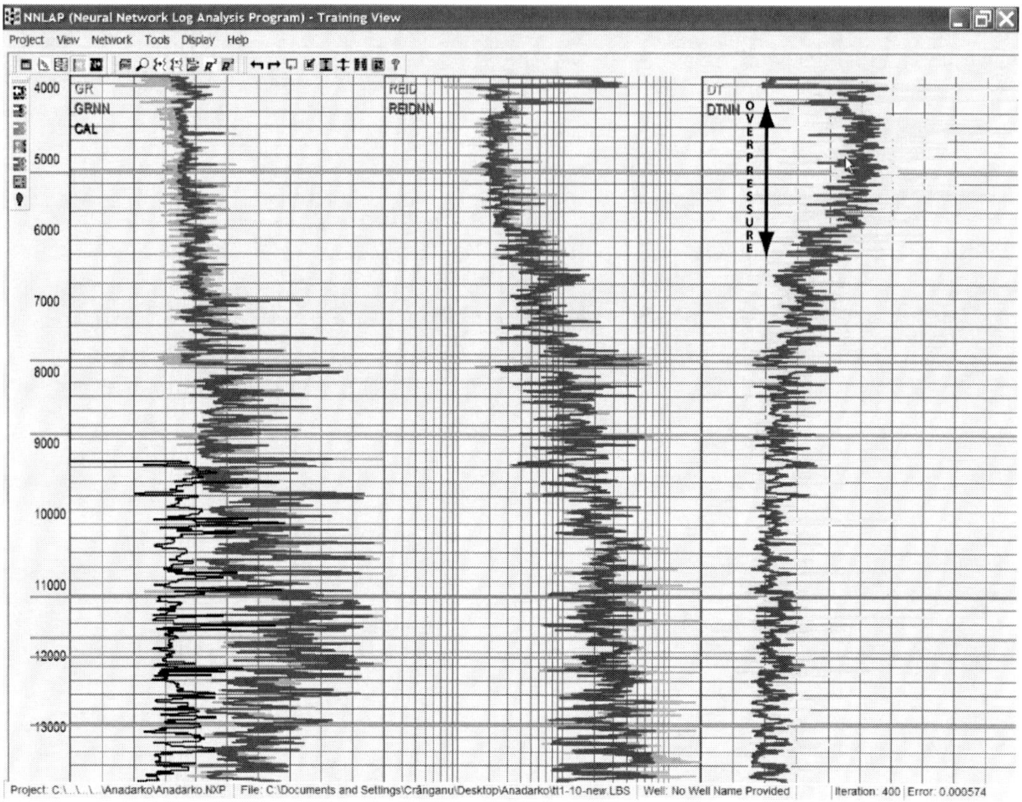

Figure 5. The training well (T& T 1-10, 35.268 N, -99.197 W). The black arrow indicates the position and extent of overpressured zone.

Using crossplots of the input and simulated data during the training time, the user can detect anomalous neural network behavior. For example, Figure 6 shows a crossplot of DT and DTN1 in the training well between 11,696 and 13,060 ft depth interval. The samples are color-coded by depth for easy identification. If samples below 11,969 ft of depth are more or less correlated, the samples colored light blue (between 11,696 and 11,969 ft of depth) show a clear lack of correlation. The user has now the opportunity to fix this problem by revising the train-ing examples or other training and design settings (e.g., number of samples in the depth inter-val)

Step 2. Confirmation and validation of the model

The back-predicted input curves reflect how well the input data is being used in the prediction of the target value in the training well and can be used to verify the model in confirmation and application wells. For these steps to be performed with high confidence, the user must ensure that all rock types, expected to be found in confirmation and application wells, are represented in the training well. If all rock types cannot be represented in a single training well more than one well can be used to train the neural network. A careful interpretation of the available data (well logs, cores, drilling reports, etc.) is necessary.

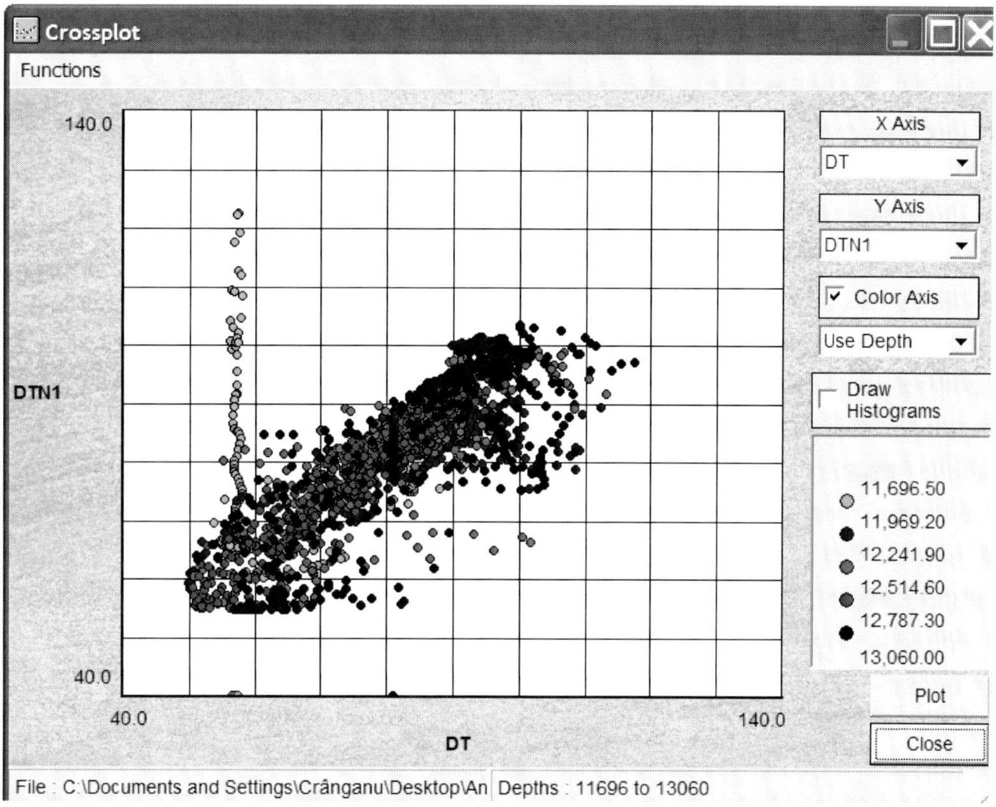

Figure 6. Crossplot of DT-DTN1 (depth interval 11,969 ft – 13,060 ft) in the training well.

Confirmation and validation of the model was performed by applying the modeled solution obtained in the training well to additional sets of data that contain the same inputs (in our examples, gamma-ray GR and resistivity REID logs) and the target values (sonic log DT). During this step, the solution from the training well can be validated through back-prediction of the predictor data and the fit between the predicted DT and the recorded DT in the confirmation well. Figure 7 shows one of the confirmation wells (Whittenberg 3-29) used to confirm the trained neural network. The confirmation (match) between yellow, recorded DT curve and the red, synthetic DT curve is considered to be accurate and the matching error is low enough to satisfy the user.

Step 3. Application of the model

Finally, the confirmed model is used to estimate the sonic log values DT in application wells containing only the input curves (GR, REID). The output is the simulated (synthetic) DT. Figures 8 and 9 show two of the application wells (Smith 1-13 and Ledbetter 1-18, respectively) in which we performed a neural network analysis to estimate the variation of sonic velocities with depth.

1.4. Predicting the Presence of Overpressured Zones in the Anadarko Basin, Oklahoma

Determining trends in acoustic transit time on sonic logs (DT) is a conventional method of predicting abnormal pressure in shales or sandstones interbedded with shales (e.g., Hottman and Johnson, 1965; Magara, 1976; Powley, 1982; Surdam et al., 1997; Hsu and Belaud, 1998).

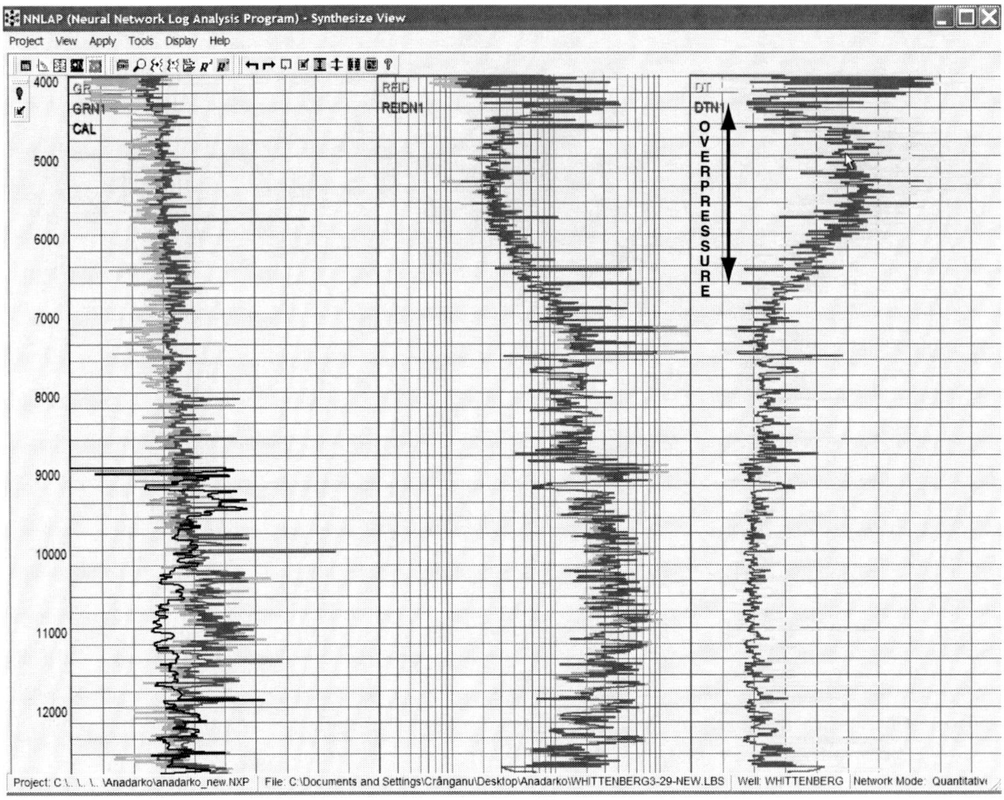

Figure 7. The confirmation and validation well (Whittenberg 3-29, 35.316 N, -99.342W). The black arrow indicates the position and extent of overpressured zone.

Because the velocity of a compressional wave in sedimentary rock is dependent on the effective rock stress, an increase in transit time is interpreted as indicating a lower effective rock stress and, consequently, overpressuring. Hermanrud et al. (1998) found that the sonic and resistivity log-derived porosities are significantly higher in the overpressured than in the normally pres-sured wells. Similarly, Poston and Berg (2002) state that sonic logs show an increase in inter-val transit times in overpressured shales. The calculation of abnormal pressure from log data is based on the relationship between observed resistivity or acoustic values of shales and a trend line reflecting normal compaction (Heppard et al., 1998)

These considerations have led us to predict the presence of overpressured zones in the training well T&T 1-10, between ~4,200 ft – 6,400 ft (1,280 m – 1,950 m) (Figure 5, see the black arrow), in the verification well Whittenberg 3-29, between ~4,400 ft – 6,400 ft (1,340 m – 1,950 m) (Figure 7, see the black arrow), as well as in the two application wells chosen for this article: Smith 1-13 well, between ~5,000 ft – 7,000 ft (1,524 m – 2,134 m) (Figure 8, see the black arrow) and Ledbetter 1-18 well, between 5,800 ft – 7,200 ft (1,768 m – 2,195 m) (Figure 9, see the black arrow). These depth intervals are mostly sandy shales with intercalations of shaly sandstones (Cranganu, 2005b). The presence of such overpressured zones (compart-ments) within the Anadarko Basin is well documented (Al-Shaieb et al., 1990, 1997a, 1997b, Cranganu, 2004).

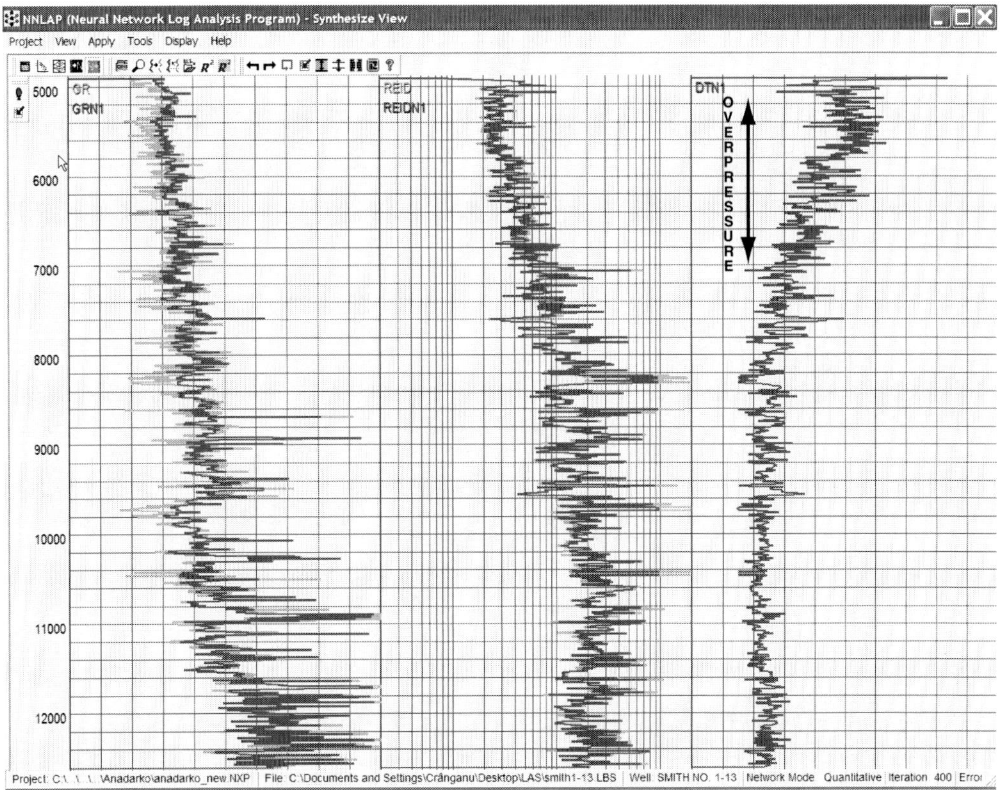

Figure 8. Application well 1 (Smith 1-13, 35.341N, -99.162W). The black arrow indicates the position and extent of overpressured zone.

1.5. Conclusion about Using ANN

Wireline log analysis is still one of the most effective methods available to estimate pore fluid pressure. It is used to create models of pressure in wells during the planning of drilling programs. Many methods of pore fluid pressure estimation, such as those using resistivity, density, or sonic logs, require, in the first instance, that the necessary logs be present for inter-pretation. Because the sonic log, considered to be accurate and sensitive to changes in the rock stress regime (e.g., overpressure), is rarely recorded in wireline logging operations, we used an artificial neural network to produce synthetic sonic logs curves based on mathematical rela-tionships obtained from the commonly available log curves. Artificial neural networks can use commonly available logs, such as normalized natural gamma (GR) and deep resistivity (REID) to created synthetic DT logs. The effective use of the ANN requires three steps: (1) supervised training of the neural network; (2) confirmation of the model validity by blind testing the solu-tion in other wells that contain the predictor curves used in developing the model from the training well (GR and REID) and a recorded compressional acoustic log (DT), and (3) validation (through back-prediction of the input logs) and application of the model to wells contain-ing only the input curves (GR, REID) in order to obtain the synthetic DT log.

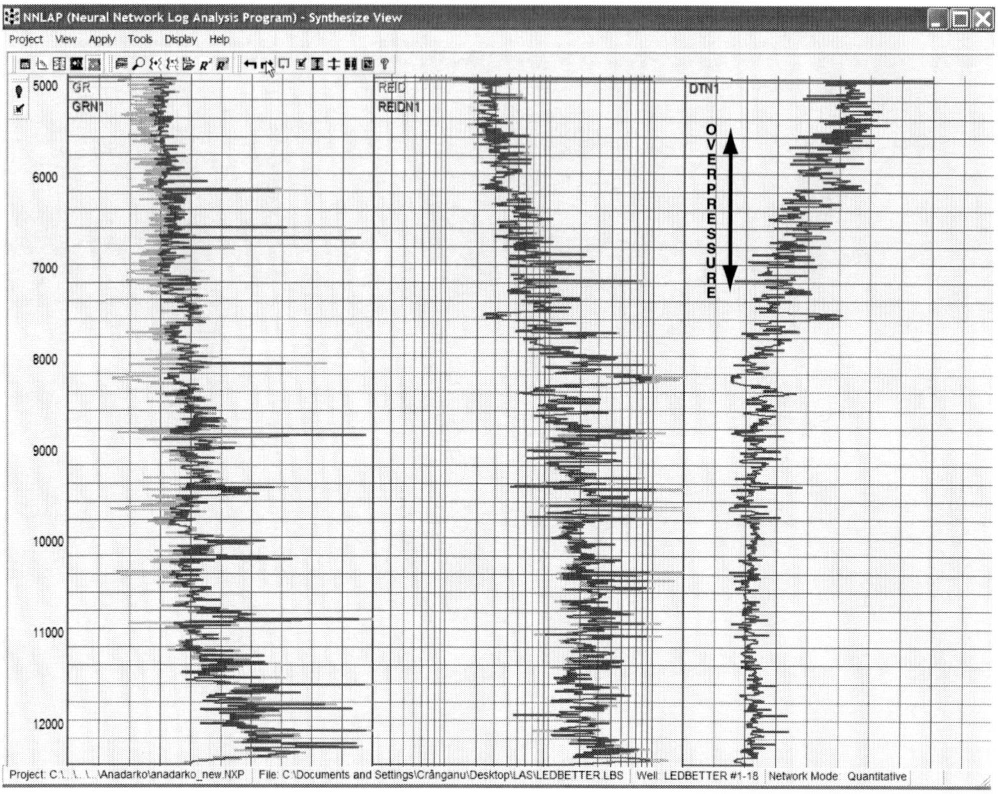

Figure 9. Application well 2 (Ledbetter 1-18, 35.341N, -99.144W). The black arrow indicates the position and extent of overpressured zone.

During the training of the artificial neural network proposed for the Anadarko Basin we found that an accurate selection of a particular log as the network input curve was not necessary because using multiple logs seems to give equivalent results as when using the highest correlated log. We agree with Saggaf and Nebrija (2003) that a redundant input containing several logs may yield reasonably accurate results as long as some of the logs in the input are sufficiently correlated with the missing log.

Using existing and simulated DT values we predicted the presence of overpressured zones in the Anadarko Basin at depths starting between 4,200 – 5,000 ft (1,280 - 1,524 m) and ending at depths between 6,400 – 7,200 ft (1,950- 2,195 m). These findings are in agreement with previous similar results (Al-Shaieb at al., 1990, 1997a, 1997b, Cranganu, 2004)

Using artificial neural networks to predict the presence of overpressured zones in the Anadarko Basin, Oklahoma, has several advantages over traditional deterministic methods. They do not require a precise mathematical model equation that describes the dependency between the predictor values and the target values and, unlike linear regression techniques, neural network methods do not over-predict mean values and thereby preserve original data variability. One of their most important advantages is that their predictions can be validated and confirmed through back-prediction of the input data.

The method presented here may be applied in other areas of the Anadarko Basin or in similar basins experiencing abnormal pore fluid pressures and where DT (sonic log) values are scarce or missing. The initial results described here are promising and encouraging for the future use and refinement of the method.

2. Gene Expression Programming (GEP)[5]

In the oil and gas industry, characterization of pore-fluid pressures and rock lithology, along with estimation of porosity, permeability, fluid saturation and other physical properties is of crucial importance for successful exploration and exploitation. Along with other well logging methods, the compressional acoustic (sonic) log (DT) is often used as a predictor because it responds to changes in porosity or compaction and, in turn, DT data are used to estimate formation porosity, to map abnormal pore-fluid pressure, or to perform petrophysical studies. Unfortunately, despite its intrinsic value, the sonic log is not routinely recorded during well logging. Here we discuss the use of a soft computing method – Gene Expression Programming (GEP) – to synthesize missing DT logs when some more common logs (such as natural gamma ray – GR, or deep resistivity – REID) are present. The Gene Expression Programming approach can be divided into three steps: (1) supervised training of the model; (2) confirmation and validation of the model by blind-testing the results in wells containing both the predictor (GR, REID) and the target (DT) values used in the supervised training; and (3) applying the predicted model to wells containing the predictor data and obtaining the synthetic (simulated) DT log. GEP methodology offers significant advantages over traditional deterministic methods. It does not require a precise mathematical model equation describing the dependency between the predictor values and the target values. Unlike linear regression techniques, GEP does not overpredict mean values and thereby preserves original data variability. GEP also deals greatly with uncertainty associated with the data, the immense size of the data and the diversity of the data type. A case study from the Anadarko Basin, Oklahoma, is presented. The results are promising and encouraging.

2.1. Introduction

Sedimentary basins containing oil and gas reservoirs are usually highly complex entities from a geological point of view. Characterization of pore-fluid pressure, rock lithology, porosity, permeability, oil or water saturation, and other physical properties is mainly based on well log data and core analysis. Each time a new borehole is drilled, a series of geophysical parameters are recorded, such as Gamma Ray (GR), Dual Induction Log, Formation Density Compensated, Deep Resistivity (REID), etc. This process is called well logging. Each such parameter varies with depth and it usually represents a specific physical characteristic of the rocks surrounding the borehole.

---

[5] This portion of our chapter draws upon the article: Cranganu, C., and Bautu, E., 2010, Using Gene Expression Programming to estimate sonic log distributions based on the natural gamma ray and deep resistivity logs: A case study from the Anadarko Basin, Oklahoma, J. Pet. Sci. and Eng. 70, 243–255. The article was a finalist for ENI Awards 2012.

Among these various well logs, the compressional acoustic or sonic log (DT) has largely been used to predict rock porosity (so called “acoustic porosity”, Ellis and Singer, 2007), to perform petrophysical analyses or to characterize areas with abnormal pore fluid via effective rock stress estimations (Cranganu, 2004, 2007; Hearst et al., 2000).

In spite of its intrinsic value for estimating rock porosity or studying and analyzing abnormal fluid pressures in sedimentary basins containing oil and gas reservoirs, the sonic log is not commonly available in oil and gas wells. This may happen either because a full suite of logs is not recorded or because of problems encountered in repeated logging, such as damaged, faulty logging instruments, or poor logging conditions in any of the logging runs. As a result, some wells may lack the sonic log entirely or the log is only partially recorded.

We discuss using a soft computing method – gene expressing programming (GEP) – to accurately synthesize missing DT data and thus more effectively map porosity variations and changes in pore-fluid pressure and increase the density of the control data in wells without recorded DT.

Application of artificial intelligence or soft computing in log analysis is bound to play a key role. Why? This is in part due to nature of the rules governing many phenomena in well logging. The uncertainty associated with the data, the immense size of the data to deal with and the diversity of the data type represent important factors to rely on unconventional tools such as soft computing (Nikravesh and Aminzadeh, 2003). Artificial intelligence techniques, such as neural computing, fuzzy reasoning, or evolutionary computing for data analysis and interpretation have become increasingly powerful tools in oil and gas industry for lowering exploration risk, reducing exploration and production costs, improving recovery through more efficient production, and extending the life of production wells.

The data sets used in the present study involve the following geophysical rock parameters: the Gamma Ray (GR) which represents the natural gamma radiation, the deep resistivity log (REID) which represents the variation of the electric resistivity, and the compressional acoustic (sonic) log (DT) which records the variation of the seismic wave propagation velocity. The caliper log (CAL), which measures the borehole diameter variations, is also available and it serves as an indicator for the quality of the recordings of the other measured parameters. We detail the procedure of selecting the data based on CAL log in the section devoted to the experiments (Section 2.5).

The problem we tackle concerns the identification of a model that best explains DT (target) as a function of GR and REID (predictors). The purpose is to find a mathematical expression that describes as well as possible the nature of the relationship between the three physical parameters of the borehole. Further, the model is to be used for predicting values of DT in situations when only GR and REID are known, and DT is unknown. Therefore, the objectives of the present study are two-fold. We are interested in establishing whether there exists a dependency between the variables GR, REID, and DT, as well as to identify a suited functional explanation for this relationship and to test its validity. The function derived is validated blindly in a borehole containing the three logs (GR, REID, and DT). The accuracy obtained by the models by comparison on wells where all three parameters are known is good. We then apply the obtained models on data recorded in two other wells, where no DT values are available. Thus, we finally obtain a synthetic DT log.

The method presented here is not limited to modeling DT values only. The algorithm used in this research can be applied (with appropriate modifications) in any area of well logging studies, where missing log values are needed.

2.2. Problem Description and Related Work

The three geophysical curves (GR, REID, and DT) are intrinsically linked, since each of them reflects some physical property of the same rock layer. For reasons mentioned above, usual well logging includes mainly the GR and the REID logs, while the third one – the sonic log (DT) – is not a part of the standard measurements.

Nevertheless, the sonic log has some very interesting applications in the study of sedimentary basins containing oil and gas reservoirs. Based on DT recordings, one can estimate rock porosity (so called "acoustic porosity") which is then used in reservoir characterization and reserve estimation (Ellis and Singer, 2007).

Combinations of sonic log with neutron log or density log represent valuable petrophysical techniques providing most of the subsurface data available to an exploration geologist. Besides their importance in well completion decisions, those techniques are also important methods for mapping and identifying lithologies (Asquith and Krygowski, 2004).

Pore-fluid pressure measured in sedimentary basins may be either higher or lower than the normal hydrostatic pressure at that depth. The existence of such abnormal regimes of pressure has implications in the evolution of hydrocarbon provinces. The sonic log (DT) has been used to characterize such areas with abnormal pore-fluid pressure via effective rock stress estimations. In this case, the DT log seems to be preferred due to its accuracy and sensitivity to changes in rock stress produced by abnormal pore-fluid pressure, especially in shale layers. Determining trends in acoustic transit time on sonic logs has become a conventional method of detecting anomalous pressures in shales (Cranganu, 2004, 2007; Hearst et al., 2000).

The problem we are dealing with is actually an inverse problem, since it consists in finding a mathematical model (explanation) for DT that reproduces a set of given observations based on known values of GR and REID. According to the criteria set by Hadamard (1902) for a problem to be well-posed, the mathematical model of a physical phenomenon must respect the following: (1) it must have a solution (existence), (2) the solution is unique, and (3) the solution is stable (it depends continuously on the data). Like many problems stemming from measurements, our problem is also a (very) ill-posed problem. It might not have a solution in the strict sense. Also, if we do encounter a perfect solution, it might not be unique, since many mathematical expressions may give out the same output. Moreover, the behavior of DT does not depend linearly on the data, as it will be shown in the experimental results section. These features make the problem a perfect candidate for heuristic approaches, such as evolutionary computation.

Previous studies trying to synthesize missing DT values used another soft computing method – Artificial Neural Networks (Cranganu, 2005, 2007). In this chapter, we employed an evolutionary algorithm to synthesize a model for the variation of the sonic log with depth, depending on series of measurements of the gamma-ray and deep resistivity.

Evolutionary algorithms were used to search for the reservoir description that matches measurements performed on the reservoir (Romero and Carter, 2001, 2003). Every part of the reservoir was described by means of Genetic Algorithm (GA) in order to characterize the flow, and the results obtained by GA presented in Romero and Carter (2003) outperformed those obtained by a Simulated Annealing algorithm. A comprehensive study regarding the use of evolutionary algorithms in the oil industry is presented by Velez-Langs (2005), particularly in the exploration and production businesses. Saemi et al. (2007) used a hybrid technique, which employed a genetic algorithm to optimize the design of a neural network in the study

the phenotype is obtained through a translation process from the genotype. This way, an important advantage of GEP over classical GP is this clear separation between phenotype and genotype (Ferreira, 2001). In GEP, the genotypes undergo selection and application of genetic operators. The fitness function evaluates the phenotype behavior with respect to its environment – the problem data set, and decides how fit the individual is.

GEP individuals are multigenic chromosomes represented by fixed length linear string of symbols. Each gene is divided into head and tail sections. The head section contains symbols from a function set F and a terminal set T, whereas the tail contains only terminal symbols. The symbols sets (F and T), the number of genes (n), and the size of the head section (h) are parameters of the algorithm. The size of the tail section is:

$$t = h(a - 1) + 1 \qquad (5)$$

where $a = \max\{arity(f) | f \in F\}$.

For example, if the head size is $h = 5$, the function set contains the common operators $F = \{*, /, -, +\}$, and the terminal set is $T = \{x, y\}$, then $a = 2$ and $t = 6$; thus, each gene contains 11 symbols. One such gene is presented in Figure 11. This gene encodes the expression tree from Figure 10. The symbols that actively participate in the phenotype form the Open Reading Frame (ORF) of the gene and are shown underlined in the figure. The rest of the symbols in the individual may become active in case some transformations in the head of the chromosome activate them (for example, if all symbols in the head were functional symbols of maximum arity). The encoding of the tree using breadth first traversal followed by padding neutral symbols up to a certain length is called Karva language (Ferreira, 2001).

The decoding process of a GEP individual to the corresponding mathematical solution is called translation. To decode a gene into an expression tree, we need to consider the arity of each symbol (variables and numerical constants are considered 0-arity) and associate each symbol with the parameters necessary for its evaluation. In the case of the gene from Figure 11, the first symbol (+) requires two parameters (the second and the third symbols, * and +); the second symbol (*) requires two parameters (2 and x); the third symbol requires two parameters (y and 3); the fourth symbol (2) requires no parameters, and so on. Therefore, by translation of the gene we obtain the following expression: $(2x) + (y+3)$. When decoding a multigenic individual, the expressions resulted from its genes are linked together by some linking function (usually addition).

GEP individuals exhibit the simplicity of the classical linear GA individuals, while maintaining the complexity of the hierarchical structures in the traditional GP individuals. This representation allows GEP to cross the phenotype threshold, as do complex natural systems (Ferreira, 2006). It allows representation of any conceivable expression, at the same time allowing a wealth of powerful genetic operators to be defined for a thorough search. Moreover, the set of GEP operators always produces valid expressions.

2.4.2. Genetic Operators

The objective of the genetic operators is to adapt genetic information from existing individuals and combine it with new information in order to create individuals which encode better solutions for the problem tackled. The GEP algorithm offers many types of genetic operators that have different intensification/diversification search properties.

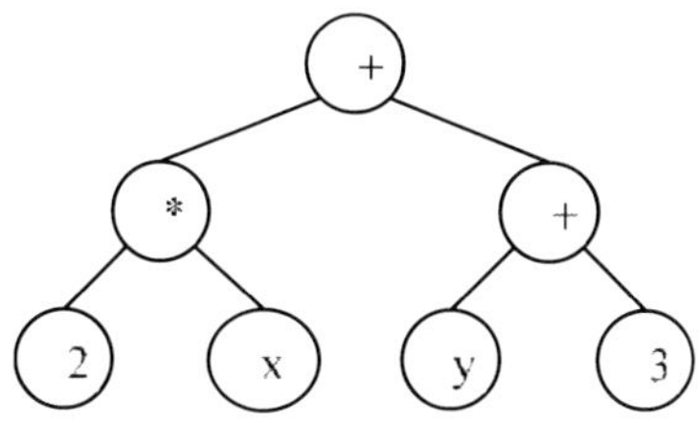

Figure 10. Expression tree of a GP individual.

```
+ * + 2 x y 3 x 5 y
        2
```

Figure 11. GEP gene for the expression tree from Figure 10 (the ORF is underlined).

The main types of GEP operators are *mutation*, *transposition*, and *crossover*. Due to the general data structure employed by GEP individuals, new operators, or operators that use problem-specific knowledge can be easily defined. The mutation operator can change any symbol in a chromosome as long as its structural organization is preserved: in the head, any symbol can change into any other symbol (function or terminal), regardless of its arity; in the tail, terminals can only change into terminals. For example, by changing the third symbol, the gene from Figure 11 would mutate into the gene from Figure 12. The new gene translates into the expression $2x + 5$. The number of active symbols has changed accordingly to 5.

Mutation is not considered a main operator in GP (Koza, 1992), but Ferreira (2006) considers mutation as the most efficient operator for genetic diversity. When coupled with elitist selection, it allows populations to evolve efficiently. Thus, in our GEP experiments we use rather high mutation rates, much higher than the ones found in nature (Futuyma, 2006).

Another type of unary GEP operators is the *transposition*: insertion sequence (IS) transposition, root IS transposition, and gene transposition. These operators copy a sequence of symbols called transposon (or insertion sequence) and insert it into some location. They differ among themselves in the way they select the insertion point and the insertion sequence. The transposon is copied entirely at the target site. The sequence remains intact at its place of origin, giving rise to two duplicate sequences resident in the same chromosome. The IS transposition operator chooses a random insertion sequence and inserts it at a random location in the head of the gene, except for the first position. The rest of the symbols in the head are shifted, without passing on in the tail. For example, if the operator selects the transposon "+ 2 x" and the second symbol as the insertion point, the gene from Figure 11 changes into the one from Figure 13 and decodes into the expression $x + 3y +2$.

In a similar way, the root IS transposition (RIS) operator chooses a random insertion sequence that begins with a function and inserts it at the first position of the head of the gene (thc root of the encoded expression tree). Using the same transposon as before, the RIS transposition operator would produce the gene from Figure 14 which decodes into $2 + x$.

Gene transposition moves an entire gene in front of the first gene of the chromosome. We do not use this operator in our research because we use addition as linking function and it is commutative.

The binary *crossover* operators work by exchanging information between two individuals, which are called parents in this context. In GEP there exist one-point, two-points and gene crossover operators. Like in GAs, they differ in the way the cut points are selected.

In one-point crossover, two individuals are selected and then a single cut point is randomly selected among the symbols of the gene. The left portion of the first parent and the right portion of the second parent are joined, and vice versa. Figure 15 presents the behavior of one-point crossover when the cut point is after the third symbol.

```
+ * 5 2 x y 3 x 5 y 2
```

Figure 12. GEP gene from Figure 11 after mutation of the third symbol.

```
+ + 2 x * y 3 x 5 y 2
```

Figure 13. GEP gene from Figure 11 after IS transposition.

```
+ 2 x + * y 3 x 5 y 2
```

Figure 14. GEP gene from Figure 11 after RIS transposition.

+ * +	2 x y 3 x 5 y 2
+ 2 3	5 - x 2 5 3 1 2
+ * +	5 - x 2 5 3 1 2
+ 2 3	2 x y 3 x 5 y 2

Figure 15. Effect of one-point GEP crossover operator (parents represented above the horizontal line, offspring are represented below; the cutting point is represented by the vertical line).

In a similar fashion, the two-points crossover operator randomly selects two cut points and exchanges information only between these points. Gene crossover randomly selects single genes and exchanges them between the parent individuals.

All the genetic operators defined in GEP create valid individuals that decode to syntactically correct mathematical expressions. This allows a thorough and fruitful search of the solutions of the problem (Ferreira, 2006).

2.4.3. Selection

The selection operator simulates the process of natural selection. In nature, individuals have different chances to survive up and produce offspring, depending of their features. In genetic algorithms, selection uses the fitness of individuals to assign higher probability for surviving to the individuals that are better adapted to their environment, thus encoding better solutions for the problem at hand. The selection method used in our experiments is roulette wheel selection combined with elitist survival of the best individual from one generation to the next. Every individual has fitness proportionate chances to survive and the best individual in a generation is guaranteed to have at least one copy in the next generation.

2.4.4. Fitness Function

Every individual in the GEP population is assigned a fitness value according to how well it solves the problem. We denote by *D* the dataset we model, consisting of tuples $d_i = (DT_i, GR_i, REID_i)$. The goal of our research is to find a function *f(GR, REID)* that follows the actual recorded values of *DT* as close as possible. In order to measure the quality of each individual, GEP usually employs a simple evaluation function, like that in equation (6), which indicates how far the output of an expression tree is from the expected *DT* values, for each (*GR, REID, DT*) observation recorded in our dataset *D*.

$$eval(f) = \sum_{(GR,REID,DT)\in D} \left| f(GR,REID) - DT \right| \tag{6}$$

Another way to assess the quality of a model is by computing the mean square error:

$$MSE(f) = \frac{1}{|D|} \sum_{(GR,REID,DT)\in D} \left(f(GR,REID) - DT \right)^2 \tag{7}$$

If we consider the relative error (*RE*) of an individual *f* with respect to a fitness case d_i,

$$RE(f,d_i) = \frac{f(GR_i, REID_i) - DT_i}{DT_i} \tag{8}$$

then we may measure the quality of the model encoded by the individual by computing the mean relative error (*MRE*) by

$$MRE(f) = \frac{1}{|D|} \sum_{d_i \in D} RE(f,d_i) \tag{9}$$

2.4.5. The General GEP Algorithm

GEP follows the same search process pattern as other evolutionary algorithms. The quest for a fit solution begins with a randomly generated population of individuals. They are subject to genetic transformations. These individuals are then evaluated based on how well they fit the given data set and they get fitness values. According to their fitness, they get higher or lower chances to survive in the next generation. The process repeats with their offspring; they go through genetic operators and reproduction, until a stopping criteria is met. The stopping criteria of the algorithm can consider different parameters of the evolution process, such as a maximum number of generations, maximum number of fitness function evaluations, minimum fitness value, etc. We stop the evolution process when a certain number of generations have passed.

Considering the elements presented in the previous sections, the GEP algorithm consists of the following steps (see Figure 16):

- Create random initial population
- Evaluate the population with formula (6)
- Apply genetic operators
- Apply selection operator
- If termination criterion is not met, then go back to Step 2.

2.5. Experiments

The methodology we used in this research was divided into three steps: (1) a supervised training of the GEP algorithm, by using three logs (GR, REID, and DT) from a training well; (2) confirmation and validation of the trained model by blind-testing the results in a verification well containing both the predictor (GR, REID) and the target values (DT); (3) applying the predictive model to application wells containing only the predictor values (GR, REID) and obtaining synthetic target (DT) values.

A Case Study: The Anadarko Basin, Oklahoma

We used a case study from the Anadarko Basin, Oklahoma, because it is well known for its extensive abnormal pore-fluid pressure, mainly over-pressured compartments (Al-Shaieb et al., 1990, 1997a, 1997b; Cranganu, 2004). The rationale for synthesizing DT data in the basin is that those data can be used to increase the density of the control data and more effectively map changes in pore-fluid pressure.

The data sets used in this chapter come from the following wells (Figure 1): training well: T&T 1-10 (35.268N, -99.197W); verification well: Whittenberg 3-29 (35.316N, -99.342W); application well #1: Smith 1-13 (35.341N, -99.162W), and application well #2: Ledbetter 1-18 (35.341N, -99.144W).

The data sets from the first two wells consist of sonic log (DT, measured in μs/m), gamma-ray log (GR, measured in uAPI) and deep resistivity log (REID, measured in Ωm). The application wells contain only the GR and REID logs. The depth interval for all wells was between 1,219 – 4,267 m (4,000 – 14,000 ft). Recordings of the Caliper (CAL) log were also available at each level of depth, but this parameter was not used as an explanatory variable in the model since it has no semantic implications in deriving the model. CAL log records the variations of the borehole diameter and it is used to assess the quality of the data recorded at a given depth such that training could be avoided in zones with noisy, low quality data caused by washouts, mud cakes, non-recording, etc.

Step 1. Model Training

Based on the analysis of the CAL log, we selected seven different depth intervals that form the actual training set used to infer the models by GEP. Best recorded data comes from sequences of data where the diameter (described by CAL) is (almost) constant. Data that are suspicious (with respect to CAL) are not included in the training set. Therefore, out of the approx. 20,000 data items available for the training well, the actual training set consists of 1407 recordings in the following depth intervals: 1,570 – 1,600 m; 2,393 – 2,423 m, 2,713 – 2,743 m, 3399 – 3429 m, 3,581 – 3,612 m, 3,658 – 3,688 m, and 3,947 – 3,978 m (see Figure 17). The Pearson correlation coefficient between the observed and computed DT values for 1407 training recordings is 0.91, indicating a very strong positive correlation.

We used the inferred model to extrapolate the DT values over all the data available from the T&T 1-10 training well, consisting of about 20,000 records (by extrapolation we denote the process of obtaining the DT values for depths that were not included in the GEP training set). The Pearson correlation coefficient for the observed and computed DT is 0.79, still indicating a strong correlation between the observed and computed DT data (Figure 18), and thus proving the robustness of the GEP model obtained. Predictions on points outside the training regions prove the good generalization capabilities of GEP, due to the smooth surface of the induced solution.

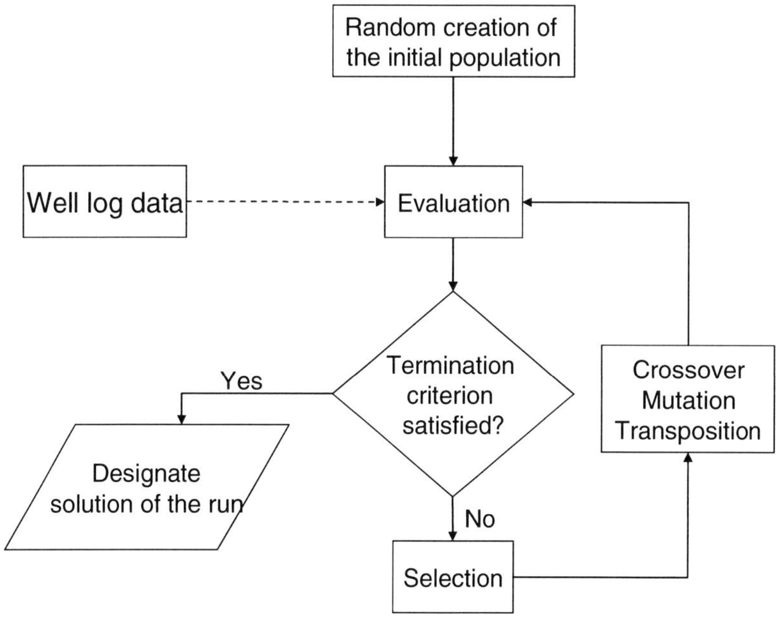

Figure 16. Flowchart of main steps in GEP.

3-D plotting of the data selected from the T&T well for the training set helps us to get a visual understanding of the ill-posedness of the problem (Figure 19). It also supports our hypothesis that there exists a dependency among the three parameters and provides motivation for continuing with the experiments. Evidence of noise in the data is provided by the existence of very close points (in terms of GR and REID) which are extremely different with respect to the expected DT values. Genetic programming is known to factor out such noise in the data and to give an overall fit solution to a given data set (Vladislavleva, 2008).

The algorithm parameters are important when using an evolutionary computation method such as GEP. The values used are as suggested in the literature (Ferreira, 2006). The experimental setup for the results presented is described next. Automatically fine-tuning of GEP parameters is possible and it may improve results obtained, but such a study is beyond the purpose of this paper. The initial population contains randomly created individuals. The solution designated by a run is the individual with the lowest mean square error over the training set obtained throughout all generations. An experiment consists in 50 independent runs with the same set of parameters, but with a different seed for the randomizer. The seeds are logged in order to ensure perfect reproducibility of the presented results. The parameters used for the GEP algorithm are presented in Table 1. The solution of an experiment is taken to be the best individual encountered throughout all the independent runs in the experiment.

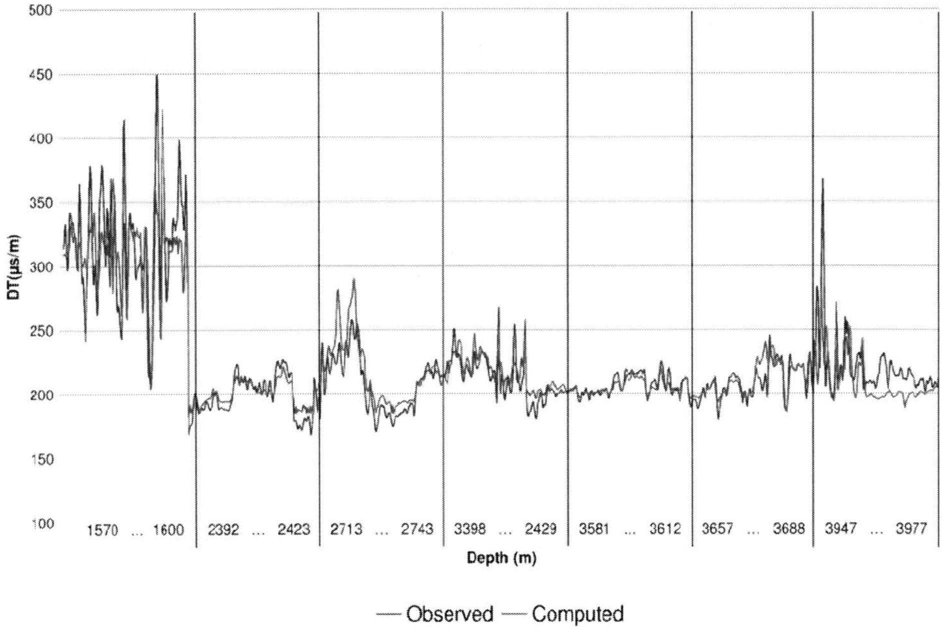

Figure 17. The observed (in green) and computed (in red) DT values in the seven training intervals from well T&T 1-10 (selected based on interpretation of caliper (CAL) log).

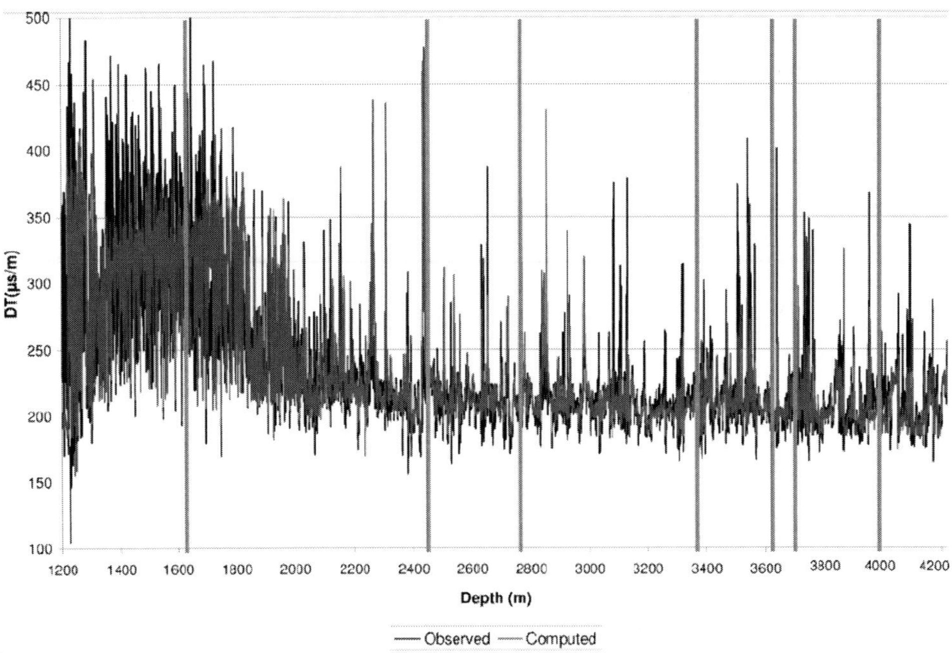

Figure 18. The observed (in black) and computed (in red) DT values in the entire training well (T&T 1-10).

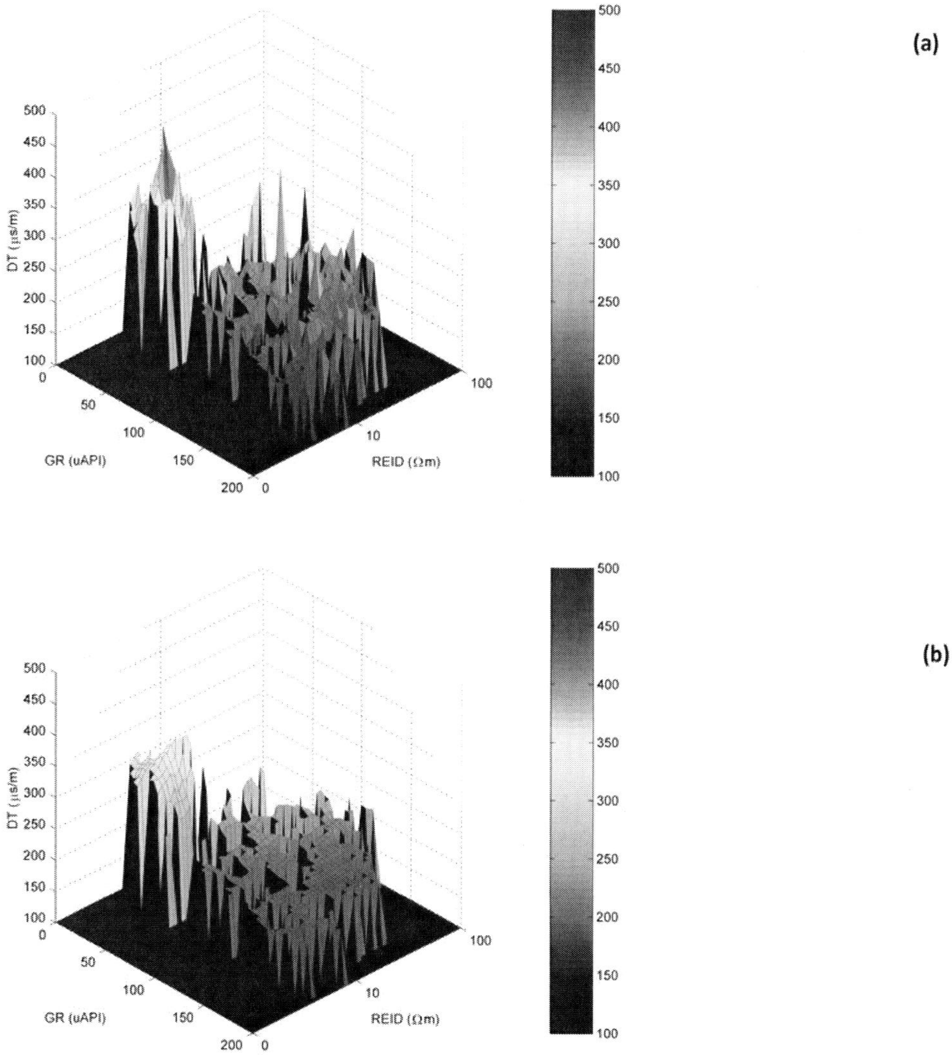

Figure 19. 3-D distribution of DT values in the training well (T&T 1-10) in correlation with GR and REID values. (a) Observed DT values; (b) Computed DT values.

The function set we used is composed of basic mathematical functions, namely arithmetic operators (+, -, *, /), natural logarithm, sine, cosine and tangent, logarithm and exponential. The choice of these specific functions was rather intuitive and it was guided by recommendations made in Ferreira (2006) and by repeated experimentation on our part. We used protected versions of the division and the logarithm operations (Koza, 1992): they will return 1 whenever the operation is not defined (e.g. division by 0).

The best solution encountered over the 50 independent runs (500 generations each) has a MSE of 30.57 on the training data set (Table 2).

By comparison, the solution that had the lowest MSE on the verification well (98.50) did not obtain the lowest MSE on the training set. Its MSE over the training data set is 31.78, slightly larger than the designated solution.

Mean and standard deviation of the MSE of all solutions from the 50 runs are reported, to account for the reliability of the method (Table 3). We also report the median and the interquartile range of the MSE recorded in the 50 runs as a measure of the performance of different runs.

Table 1. Parameters used for GEP algorithm

Terminal set	GR, REID Random constants in [-20, 20]
Function set	+, -, *, /, sin, cos, log, exp
No. of generations	500
Population size	50
Fitness function	MSE
No. of genes	5
Head size	5
Linking function	Addition (+)
Mutation rate	0.03
IS Transposition rate	0.1
RIS Transposition rate	0.1
One Point Crossover rate	0.3
Two Points Crossover rate	0.3
Gene Crossover rate	0.1

Table 2. Results of solution of 50 independent runs

Data set	MSE	MRE
Training well (T&T 1-10)	30.57	0.045
Verification well (Whittenberg 3-29)	106.17	0.077

Table 3. Statistics over all the 50 solutions designated in the independent runs of the experiment

Data set	Mean MSE	Standard deviation of the MSE	Median MSE	IQR
Training well (T&T 1-10)	44.12	6.47	46.25	4.07
Verification well (Whittenberg 3-29)	113.00	14.52	110.89	8.58

Step 2. Model Verification

The model obtained by training on selected data from the training well was next confirmed and validated by blind-testing on the data set from the verification well Whittenberg 3-29, which consists of approximately 8900 records. The observed 3-D distribution of DT values is shown in Figure 20 along with the distribution of DT values computed by the GEP model. The model learned from the training well obtained on the verification well a MSE value of 106.17. Nevertheless, the MRE obtained on the same data set is 0.077, which attests that the model is actually quite good at fitting data from this unknown well.

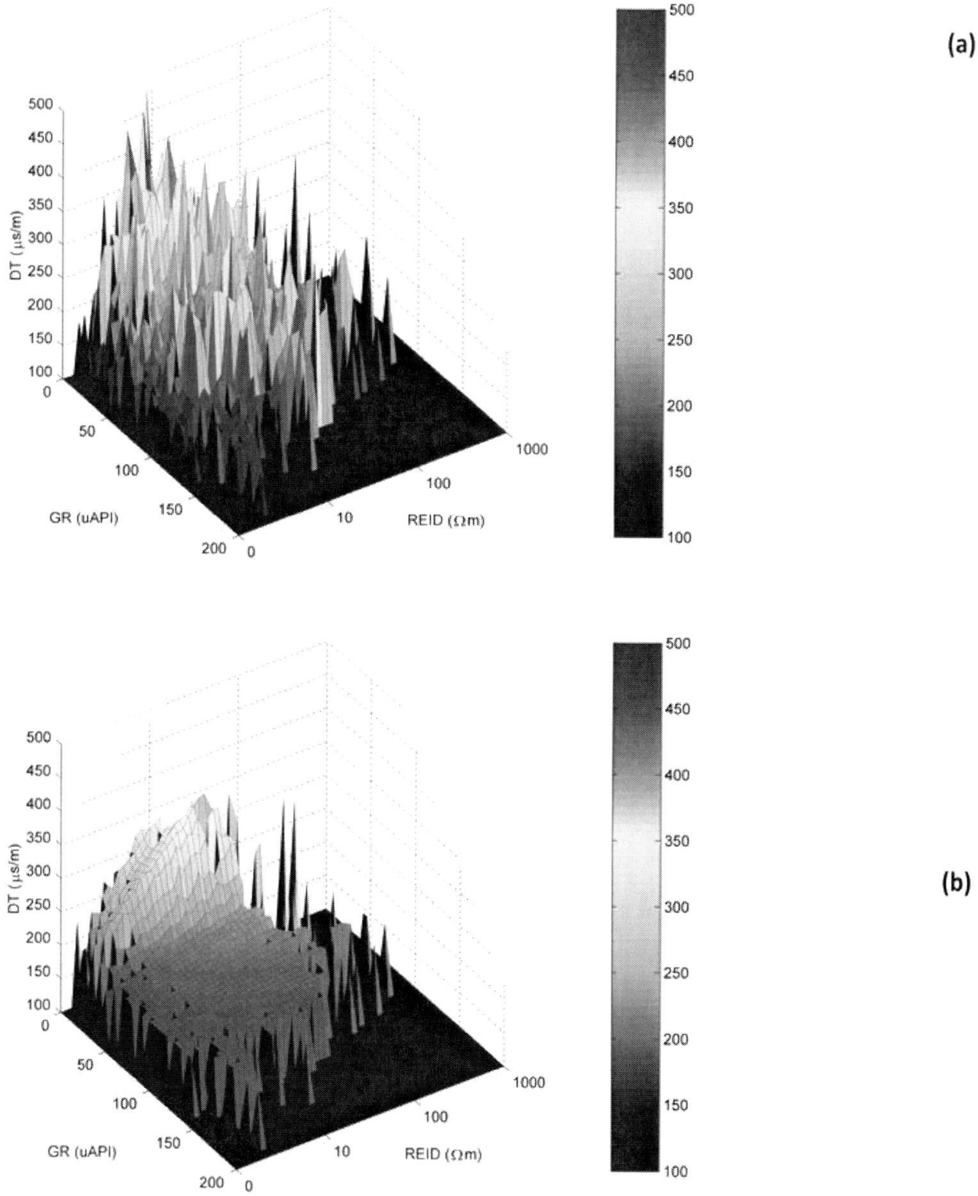

Figure 20. 3-D distribution of DT values in the verification well (Whittenberg 3-29) in correlation with GR and REID values. (a) Observed DT values; (b) Computed DT values.

The correlation coefficient on the verification well is 0.81, a very satisfactory value that ensures the validity of the obtained model. This proves that it is possible to predict DT values from the verification well using the model learned on the training well (i.e. the model is consistent).

The mathematical formula obtained by decoding the solution developed by GEP is presented next for exemplification purposes only:

$$f(GR, REID) = \sin\frac{GR}{GR - REID} + e^4 + e^{\tan(\sin\frac{-12}{REID})} + \frac{GR}{\cos(REID) + REID} + \cos\frac{15}{REID} \tag{10}$$

Step 3. Model Application

The final stage in our experiment consists in predicting values of the sonic log for two wells with no available measurements of this parameter. On these two wells (Figures 21 and 22), the values predicted by the model cannot be verified or validated, but can only be analyzed for consistency with the knowledge gained from the first two wells. As a rule, we analyzed many solutions found by the algorithm and decided to keep the one which produced the smallest MSE during training.

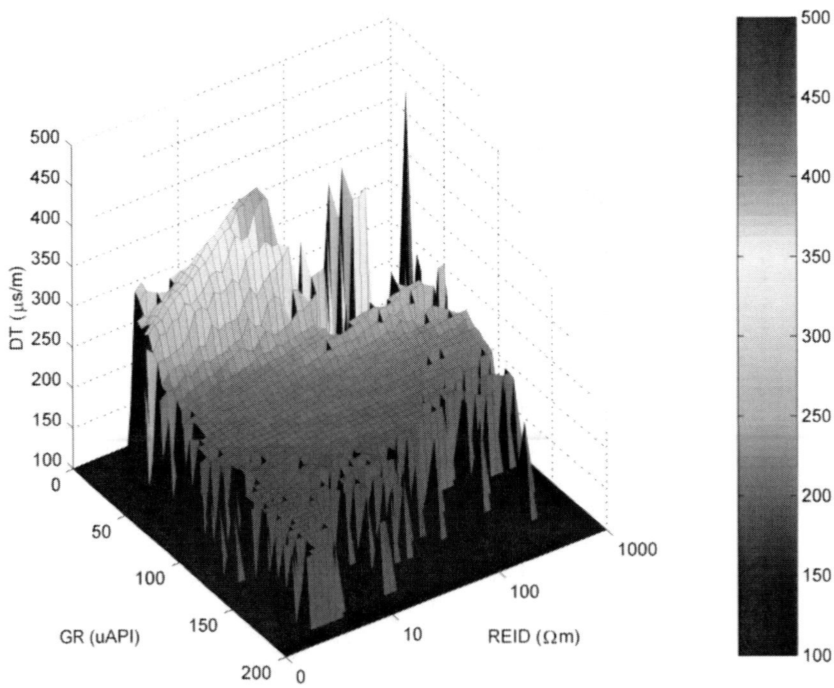

Figure 21. 3-D distribution of computed DT values in the application well #1(Smith 1-13), in correlation with GR and REID values.

2.6. Conclusion about Using GEP

We used a soft computing method, the Gene Expression Programming, to simulate missing acoustic logs in wells where other logs (e.g., gamma ray, deep resistivity) are available. Our approach involved three steps: (1) supervised training of the model; (2) confirmation and validation of the model by blind-testing the results in wells containing both the predictor (GR, REID) and the target (DT) values used in the supervised training; and (3) applying the predicted model to wells containing the predictor data and obtaining the synthetic (simulated) DT log.

GEP methodology offers significant advantages over traditional deterministic methods. It does not require a precise mathematical model equation describing the dependency between the predictor values and the target values. Unlike linear regression techniques, GEP does not overpredict mean values and thereby preserves original data variability. GEP also deals greatly with uncertainty associated with the data, the immense size of the data and the diversity of the data type.

The experiments reported in this study used large real world data sets and the solution encountered may unveil relationships hidden in the data that are crucial for understanding the data source. Although the models we derive are purely empirical, the large amount of data and the experimental methodology we used provide a high level of confidence for the validity of the symbolic model we obtained.

The functional expressions evolved by GEP are quite complex and they may be difficult to understand, even after symbolic post processing with tools like Mathematica (Kirshenbaum and Suermondt, 2004). The evolutionary approach provides multiple diverse empirical models without the need for additional knowledge about the problem. It does not enforce a specific functional form of the model and provides analytical alternatives to numerical solutions obtained by other methods. The robust model obtained provides reliable predictions outside the training range; the smooth surfaces induced by GEP minimize the potential for erroneous responses when using the model on data not used in the training process.

Other than basic statistical tests of the performance of the GEP algorithm and visual inspection of results, no convergence or stability result can be formulated, mainly due to the stochastic nature of all the algorithms in the evolutionary computation domain. Theoretical studies are very scarce in the GP/GEP field and are an open area for research.

A case study involving sets of wireline logs from four wells in the Anadarko Basin, Oklahoma, exemplified our method. The robustness of the model, as estimated by statistical parameters, such as Pearson correlation coefficient, mean square error, and relative mean error, was found to be good. The method studied in this paper can be extended in other domains of simulating missing data and the results obtained are promising and encouraging.

3. Support Vector Regression[6]

Oil and gas exploration in sedimentary basins containing highly complex geologic formations requires, among others, characterization of pore-fluid pressure, rock lithology, porosity, permeability, oil or water saturation, and other physical properties. An important tool to help solving these problems is well logging, during which several geophysical

---

[6] This portion of our chapter draws upon the article: Cranganu, C., and Breaban, M., 2013, Using support vector regression to estimate sonic log distributions: A case study from the Anadarko Basin, Oklahoma, J. Pet. Sci. and Eng. 103, 1-13.

parameters, such as Gamma Ray (GR), Dual Induction Log, Formation Density Compensated, Deep Resistivity (REID), Self-Potential (SP[7]), etc., are recorded.

Each such parameter depends on a specific physical characteristic of the rocks surrounding the borehole (e.g., electric resistivity, natural radioactivity content, electric potential, etc.) and its variation with depth is recorded as a well log. Among various well logs, the compressional acoustic or sonic log (DT) has largely been used to predict rock porosity (so called "acoustic porosity", Ellis and Singer, 2007), to perform petrophysical analyses or to characterize zones with abnormal pore-fluid pressures via effective rock stress estimations (Cranganu, 2005a, 2007; Hearst et al., 2000, Cranganu and Bautu, 2010).

Although the sonic log has proved many times its efficiency in estimating rock porosity or studying and analyzing abnormal pore-fluid pressures in sedimentary basins containing oil and gas reservoirs, it may not be commonly available in well logging suites. The reasons include lacking a full suite of logs, or problems related to damaged, faulty logging instruments, or poor quality recordings. Consequently, some wells may lack the sonic log entirely or the log is only partially recorded.

We discuss using a method belonging to the class of supervised machine learning algorithms — Support Vector Regression (SVR) — to synthesize missing compressional acoustic or sonic (DT) logs when only common logs (e.g., natural gamma ray — GR, or deep resistivity —REID) are available.

The sonic log is often used as a predictor of pore-fluid pressure changes because it responds to changes in porosity or compaction produced by abnormal pore-fluid pressures. A case study from the Anadarko Basin, Oklahoma, involving about estimating the presence of abnormally pressurized pore-fluid overpressured zones by using synthesized DT values, is presented Application of artificial intelligent methods in well logging analysis is going to play a key role. This is in part due to specific rules governing many phenomena in well logging. The uncertainty associated with the data, the immense size of the data to deal with and the diversity of the data type represent important factors to rely on unconventional tools such as artificial intelligent techniques (Nikravesh and Aminzadeh, 2003). Such techniques, including neural computing, fuzzy reasoning, or evolutionary computing for data analysis and interpretation have become increasingly powerful tools in oil and gas industry for lowering exploration risk, reducing exploration and production costs, improving recovery through more efficient production, and extending the life of production wells (Cranganu and Bautu, 2010).

The main data sets used in the present study include the following geophysical rock parameters: the Gamma Ray (GR), which represents the natural gamma radiation, the deep resistivity log (REID), which represents the variation of the electric resistivity, and the compressional acoustic (sonic) log (DT) which records the variation of the seismic wave propagation velocity. Other data sets include the Density (DEN[8]), which records the density variation with depth in the borehole, Self-Potential (SP), a measurement of natural electric potential, and Photoelectric Factor (PE[9]), which is a reliable indicator of borehole lithology. The caliper log (CAL), which measures the borehole diameter variations, is also available and it serves as an indicator for the quality of the recordings of the other measured parameters.

[7] Natural electric potential, in mV.

[8] Formation density, in g/cm^3.

[9] Estimator of formation lithology, unitless, but usually in barn/e.

We detail the procedure of selecting the data based on CAL log in the section devoted to the experiments (Section 3.5).

The major goal of using this method is to identify a mathematical model that best explains DT (target) as a function of GR, REID, DEN, SP, and PE (predictors). Further, the model is to be used for predicting values of DT in situations when only GR and REID (two of the most common logs) are known, and DT is unknown. Consequently, the objectives of the present study are two-fold. First, we are interested in finding a possible dependency between the variables GR, REID, DEN, SP, PE, and DT, as well as a functional explanation for this relationship and in testing its validity. The function derived is validated blindly in a borehole containing the predictors and the target. Second, we apply the obtained model on data recorded in two other wells, where no DT values are available. Thus, we finally obtain a synthetic DT log.

Additionally, we will show how DT data can further be used to estimate the presence and extent of zones with overpressured pore fluids using an example from the Anadarko Basin, Oklahoma. The use of SVR in estimation problems is motivated by its good prediction power and generalization capability. However, its application for predicting sonic logs is not a trivial task. After describing the mathematical model behind SVR, this study steps through a complex experimental analysis highlighting key points, decisions which are critical for the success of the method in this field.

The methodology presented in this chapter is not limited to modeling DT values only. It can be applied (with appropriate modifications) in any area of well logging studies, where missing log values are needed.

The rest of the paper proceeds as follows. The next section gives a brief description of the problem and previous approaches to solving it, and provides bibliographic references to other works that used Support Vector Regression in geology studies. Then we describe the SVR procedure used in this paper. Next section provides experimental results. A final discussion and conclusion section concludes this chapter.

3.1. Support Vector Regression (SVR) – Problem Description and Related Work

The geophysical parameters GR, REID, DEN, SP, PE, and DT are intrinsically linked, since each of them reflects some physical property of the same rock layer. Variations in one of the logs, due to lithology, fluid content (water, oil, or gas), pore fluid pressure, structural elements, etc., will be synergistically reflected in the other two (Cranganu and Bautu, 2010). The sonic log itself has some interesting applications in the study of sedimentary basins containing oil and gas reservoirs. Based on DT recordings, one can estimate rock porosity (so called "acoustic porosity") which is then used in mapping, lithology identification, reservoir characterization and reserve estimation (Ellis and Singer, 2007).

Another important application of sonic log is studying the abnormal pore-fluid pressure in sedimentary basins. That pressure may be either higher (overpressure) or lower (underpressure) than the normal hydrostatic pressure at that depth. The existence of such abnormal regimes of pressure has implications in the evolution of hydrocarbon provinces, reservoir characterization, and drilling operations. The sonic log (DT) has been used to characterize such areas with abnormal pore-fluid pressure via effective rock stress estimations. In this case, the DT log seems to be preferred due to its accuracy and sensitivity to changes in rock stress produced by abnormal pore-fluid pressure, especially in shale layers. Determining trends in acoustic transit time on sonic logs has become a conventional method of detecting anomalous pressures in shales (Cranganu, 2004, 2007; Hearst et al., 2000; Cranganu and Bautu, 2010).

Previous studies trying to synthesize missing DT values used soft computing methods – Artificial Neural Networks (Cranganu, 2005a, 2007) and Gene Expressing Programming (Cranganu and Bautu, 2010). In this part of our chapter, we investigate the capabilities of a machine learning method, SVR. The method was previously applied to porosity prediction (Al-Anazi and Gates, 2010a) and to permeability prediction (Al-Anazi and Gates, 2010b).

SVR (Drucker et al., 1996) has its roots in Support Vector Machines (SVMs), a mathematical model developed by Vapnik and collaborators (Cortes and Vapnik, 1995; Vapnik et al., 1997) at AT&T Bell Laboratories for supervised classification analysis. SVMs have quickly become very popular, being nowadays one of the state-of-the-art classification algorithms. SVR is just an SVM for regression analysis. This chapter deals with a variant of the SVM called ε-SVR, described next.

Generally, the regression problem is stated as follows: given a set of *training* data $\{(x^{(1)}, y^{(1)}), (x^{(2)}, y^{(2)}), \ldots, (x^{(n)}, y^{(n)})\}$ where each example i consists of a vector of numerical features $x^{(i)} \in R^d$ and a numerical variable $y^{(i)} \in R$, a function (henceforth called *regression* function) $f : R^d \to R$ has to be derived such that for each example i, $f(x^{(i)}) \approx y^{(i)}$, i.e., f predicts as close as possible the $y^{(i)}$ based on the input $x^{(i)}$. Regression is generally an optimization problem and the objective/cost function is most frequently formulated as the sum (over all training examples) of squared deviations of the predicted values from the real/training values and is to be minimized:

$$\sum_{i=1}^{n}(f(x^{(i)}) - y^{(i)})^2 \tag{11}$$

In the simplest approach to regression – called linear regression, f is formulated as a linear combination of the input variables x:

$$f(x) = w \cdot x + b \tag{12}$$

where b is a constant, $w \in R^d$ is a vector of numerical weights and $w{\cdot}x$ is the dot product. The regularized linear regression problem is formulated as follows:

$$\min_{w} \frac{1}{2n}\left(\sum_{i=1}^{n}(f(x^{(i)}) - y^{(i)})^2 + C\sum_{j=1}^{d} w_j^2\right) \tag{13}$$

The second sum is introduced in order to avoid overfitting on the training data and is called the *regularization term*. C is a constant called the regularization parameter and controls the trade-off between two goals: fitting the training set well and keeping the weights w low in order to avoid overfitting.

In ε-SVR the regression function f is also formulated as a linear combination of features (Equation 12) but the search for the weights vector w is guided by a slightly different objective: no deviation of the predicted values from the real values is allowed to be larger than ε. Therefore, the sum of squared deviations in Equation (13) corresponding to the regularized linear regression is replaced by constraints and the ε-SVR problem becomes:

$$\min_{w} \frac{1}{2}\sum_{i=1}^{d} w_i^2$$
$$\text{subject to} \begin{cases} y^{(i)} - w \cdot x^{(i)} - b \leq \varepsilon \\ w \cdot x^{(i)} + b - y^{(i)} \leq \varepsilon \end{cases} \quad (14)$$

This is a constrained convex optimization problem. In practice it might be the case that this problem is infeasible. Therefore *slack variables* are introduced to allow some constraint violations – corresponding to data items for which predictions exceed the minimum imposed deviation. Under the new formulation the objective/cost function penalizes the deviations exceeding ε and the problem becomes:

$$\min_{w,\xi,\xi^*} \left(\frac{1}{2}\sum_{i=1}^{d} w_i^2 + C\sum_{j=1}^{n}(\xi_j + \xi_j^*)\right)$$
$$\text{subject to} \begin{cases} y^{(i)} - w \cdot \mathrm{x}^{(i)} - b \leq \varepsilon + \xi_i \\ w \cdot \mathrm{x}^{(i)} + b - y^{(i)} \leq \varepsilon + \xi_i^* \\ \xi_i, \xi_i^* \geq 0 \end{cases} \quad (15)$$

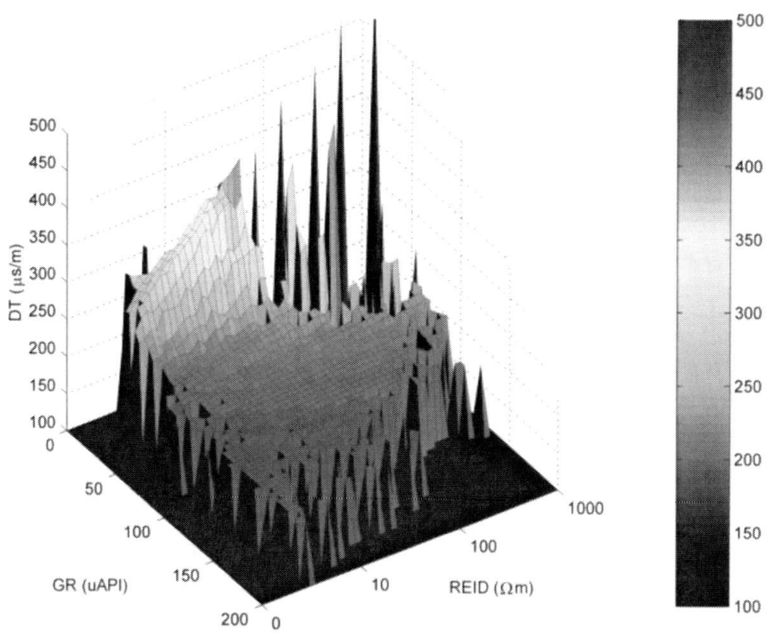

Figure 22. 3-D distribution of computed DT in the application well #2 (Ledbetter 1-18) in correlation with GR and REID values.

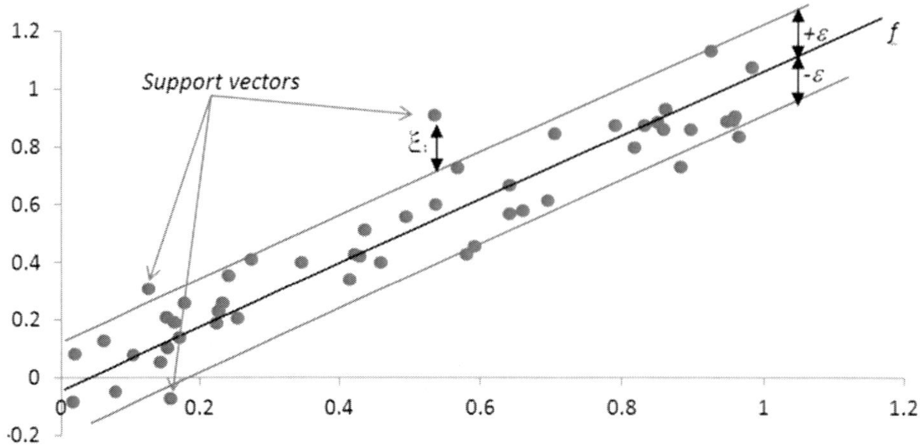

Figure 23. Illustration of ε-SVR: linear regression function *f* built by penalizing only deviations that exceed ε; the data items that deviate from *f* by a value larger than ε are called *support vectors* and their deviations from the ε-boundary correspond to the slack variables (ξ_i and ξ_i*).

Figure 23 illustrates the ε-SVR model. Only deviations larger than ε are penalized; therefore only a small number of slack variables are non-zero and are actually summed up within the objective function – those corresponding to the training items which are outside the ε tube.

3.1.1. The Dual Formulation

The problem in equation (15) is a convex optimization problem. It is usually solved in its dual formulation which provides the key for extending the linear model to generalized nonlinear functions. The constraint qualifications of the strong duality theorem are satisfied, thus the duality gap is 0 meaning that an optimal solution of the primal problem is given by the dual problem. The equivalent dual problem obtained based on the standard Lagrange multipliers method and on the saddle point condition as formulated in (Fletcher, 1989), is given in equation (16):

$$\max_{\alpha,\alpha^*}\left(-\frac{1}{2}\sum_{i,j=1}^{n}(\alpha_i-\alpha_i^*)(\alpha_j-\alpha_j^*)(x^{(i)}\cdot x^{(j)})-\epsilon\sum_{i=1}^{n}(\alpha_i+\alpha_i^*)+\sum_{i=1}^{n}y^{(i)}(\alpha_i-\alpha_i^*)\right)$$
$$\text{subject to }\sum_{i=1}^{n}(\alpha_i-\alpha_i^*)=0\text{ and }\alpha_i,\alpha_i^*\in[0,C] \qquad (16)$$

From the saddle-point condition it follows that $w=\sum_{i=1}^{n}(\alpha_i-\alpha_i^*)x^{(i)}$, i.e., w is computed as a linear combination of the input data items. The regression function becomes

$$f(x)=\sum_{i=1}^{n}(\alpha_i-\alpha_i^*)(x^{(i)}\cdot x)+b \qquad (17)$$

Based on the Karush-Kuhn-Tucker conditions (Karush, 1939; Kuhn and Tucker, 1951) that state that at the point of the solution the product between dual variables and constraints is equal to 0, some important properties can be derived:

- $\alpha_i\alpha_i^* = 0$, meaning that at least one variable is 0
- for all samples inside the ε–tube both α_i and α_i^* are 0, which means that the weight vector w and consequently the prediction function f is built only on the data items $x^{(i)}$ that lie outside the ε–tube; these data items are called *support vectors.*

3.1.2. Extension to the Nonlinear Case

The simplest way to extend linear regression towards nonlinear models is to build new features based on the existing ones. For example, by introducing polynomial features built as x_i^2 and x_ix_j, the model f becomes a quadratic one (ellipsoidal-shaped); larger exponents lead to more irregular shapes. The computational model solving the linear regression problem can be used as is under this extension. However, the complexity of this approach increases with the number of original features and the degree of the polynomial as more combinations are possible and can become overwhelming in practice. More precisely, the number of monomials of degree i that can be built from m variables is resulting in a total number of attributes that can be introduced to derive polynomial models up to a certain degree p.

Fortunately, the extension to nonlinearity in SVR is more elegant and straightforward. It comes from the observation that in both the objective/cost function (16) and the regression function (17), feature vectors x are present only in the context of dot products. Instead of computing the dot products in the enhanced feature space $\phi(x)$ where new features are added to x, it suffices to know the function $k : R^d \times R^d \rightarrow R,\ k(x^{(i)}, x^{(j)}) = \phi(x^{(i)}) \cdot \phi(x^{(j)})$. Using functions k that correspond to a dot product in some feature space $\phi(x)$, there is no need to compute the feature mapping itself. Such functions k are called *kernels*. The most popular in the field of SVR are the polynomial kernel (equation 18), the radial kernel (equation 19) and the sigmoid kernel (equation 20), respectively.

$$k_{pol}(x^{(i)}, x^{(j)}) = (x^{(i)} \cdot x^{(j)} + c)^p \tag{18}$$

$$k_{rad}(x^{(i)}, x^{(j)}) = e^{-\frac{\|x^{(i)} - x^{(j)}\|^2}{2\sigma^2}} \tag{19}$$

$$k_{sig}(x^{(i)}, x^{(j)}) = \tanh(ax^{(i)} \cdot x^{(j)} + r) \tag{20}$$

Parameter p in the polynomial kernel is called the degree of the kernel.

The radial kernel is defined by a Gaussian function. is the Euclidean distance between the two feature vectors and σ is a parameter of the radial kernel.

The sigmoid kernel has two parameters: a is a scaling parameter of the input data, and r is a shifting parameter that controls the threshold of mapping.

The SVR setting where no kernel is used is equivalent to the SVR setting employing the kernel function $k(x^{(i)}, x^{(j)}) = x^{(i)} \cdot x^{(j)}$. Because under this setting the SVR delivers a linear regression function, this kernel form is called the *linear* kernel.

The SVR optimization problem is reformulated in the nonlinear context simply by replacing each dot product by a kernel function:

$$\max_{\alpha,\alpha^*}\left(-\frac{1}{2}\sum_{i,j=1}^{n}(\alpha_i-\alpha_i^*)(\alpha_j-\alpha_j^*)k(x^{(i)},x^{(j)})-\epsilon\sum_{i=1}^{n}(\alpha_i+\alpha_i^*)+\sum_{i=1}^{n}y^{(i)}(\alpha_i-\alpha_i^*)\right)$$
$$\text{subject to } \sum_{i=1}^{n}(\alpha_i-\alpha_i^*)=0 \text{ and } \alpha_i,\alpha_i^*\in[0,C] \tag{21}$$

The regression function is given by

$$f(x)=\sum_{i=1}^{n}(\alpha_i-\alpha_i^*)k(x^{(i)},x)+b \tag{22}$$

All the kernels above give positive definite kernel matrices that guarantee that the optimization algorithm does not get trapped in local minima. The main advantage of using kernels is that nonlinear functions can be approximated with SVR while the computational efficiency of the linear approach is maintained. However, a drawback is inherent: the most popular kernel functions map the data into an infinite dimensional Hilbert space and there is not a clear and concrete understanding of what exactly these feature maps are; therefore, the nonlinear models derived by SVR are impossible to interpret in terms of relations among features/variables.

3.1.3. Parameters

In ε-SV regression two numerical parameters are essential for the behavior of the learned regression model.

The value of epsilon (ε) determines the level of accuracy of the prediction function. If epsilon is too large there will be loss in the accuracy of the model. On the other hand, if epsilon is too small the model is more prone to overfitting on the training data. Also, the complexity of the model increases with the decrease of epsilon, as more support vectors will be involved.

Parameter C acts in the opposite way. Small values determine low accuracy and underfitting. Large values introduce high penalties for vectors exceeding the epsilon threshold and may lead to complex models prone to overfitting.

Hence, both C and epsilon parameters affect the model complexity and accuracy. If kernels are used, more specific parameters come into play and have to be considered when training the model.

In order to obtain an accurate model with good generalization capability, the experimental setup described in this paper employs a search in the space of parameters defined by ε, C and the kernel-specific parameters. Two scenarios are investigated in order to address potential particularities of geological data: cross-validation that requires one data set, and the use of two data sets - one for training and one for validation.

3.2. Experiments

The experimental analysis is conducted towards finding the best model for predicting the DT parameter from log data. To this end the following features are investigated:

- The effect of normalization of the log data on the performance of the model;
- The selection of the training set and its effects on the complexity of the model;
- (SVR-specific) parameters selection;
- The predictability power of various well logging parameters.

All experiments are carried out in R using the extension of the well-known libsvm library (Chang and Lin, 2011). Specifically, the *svm* function from package 'e1071' developed by Dimitriadou et al. (2012) is used to train an ε-SV regression model with given parameters.

Following previous work of Cranganu (2005a, 2007) and Cranganu and Bautu (2010), our first experiments used SVR to predict DT based on two predictors: gamma-ray log data (GR) and deep resistivity log data (REID). After a proper setup of the experimental framework based on SVR, more well logging parameters are introduced in the analysis.

First experiments (unreported in this chapter) were designed to choose the most appropriate kernel to be used with our data sets. After trying several settings of parameters characterizing each kernel function, the results indicated the radial kernel as the most appropriate one. Therefore, all the experiments described in the following sections make use of the radial kernel.

3.2.1. Evaluation Measures

The quality of the trained model is evaluated based on the prediction accuracy. The *Mean Squared Error* (*MSE*) is computed as the average over all squared deviations of the predictions from the real values:

$$MSE(f) = \frac{1}{n}\sum_{i=1}^{n}(f(x^{(i)}) - y^{(i)})^2 \tag{23}$$

The *Mean Relative Error* (*MRE*) computes the average deviation reported to the real value:

$$MRE(f) = \frac{1}{n}\sum_{i=1}^{n}\frac{f(x^{(i)}) - y^{(i)}}{y^{(i)}} \tag{24}$$

The Pearson correlation index is also used across the paper to measure the linear correlation between the predictions and the real values.

3.2.2. Training, Validation, Testing, and Application

The steps usually involved in regression analysis consist of:

Training. In this step the regression function is computed based on a *training set* consisting of feature vectors x and an output variable y. The parameters of the method (in the case of ε-SVR parameters ε, C and kernel parameters) take fixed values.

Testing. In this step the prediction function derived in the training phase is evaluated on a test set consisting of feature vectors x and an output variable y. The test set and the training set are usually disjoint. This step evaluates the ability of the regression function to predict unseen examples. If the accuracy on unseen data is low the model is supposed to have overfitted the training data; in consequence it is not reliable for prediction purposes on data sets other than the one used for training.

Application. The regression function with a good prediction ability proved on test data is applied to an *application data set* consisting only of feature vectors x to predict the missing values for the output variable y.

Because in ε-SV regression the learned regression function depends on the values of the parameters discussed in section 3.1.3, different parameters settings in the training phase lead to different regression functions, some better than others with respect to their prediction accuracy. Therefore, an intermediate step consisting in the selection of the parameters of the method is necessary: several regression functions corresponding to various parameter settings are trained on the same training set and their prediction power is assessed on a second data set called *validation set*; the best model is considered to be the one achieving the most accurate predictions (equivalent to the lowest MSE) on the *validation set*. The validation step involving a data set different from the training one ensures that the regression function does not overfit the training set due to particular values of the parameters. The testing phase is still necessary after choosing the regression function corresponding to the best found parameters of the method: the training phase fits the weights w to the training set given an ε-SVR parameters setting, the validation phase fits the ε-SVR parameters to the validation set, and the generalization ability of the resulted regression function can be fairly assessed only by using a third data set – the test set. The importance of parameter selection and the choice of the validation method are discussed in detail in section 3.2.6.

The entire process can be described as a nonlinear multi-input vs. single-output sonic log prediction and is illustrated in Figure 24. The well logging parameters indicated in the figure are detailed next.

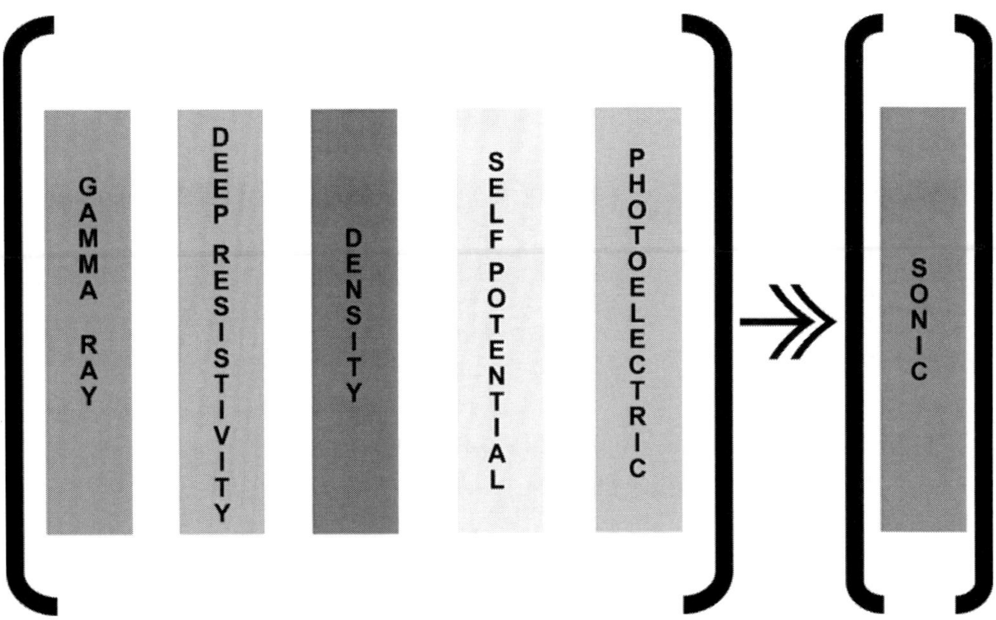

Figure 24. Example of multi-input vs. single-output sonic log prediction using ε-SVR. Details about the well logging parameters depicted in the figure are given in text.

3.2.3. Data Sets and Features

In order to validate the use of ε-SVR in the context of log data recorded in oil and gas wells, we employed data sets obtained from eight wells drilled in the Anadarko Basin, Oklahoma.

The study involves the following well logging parameters: gamma-ray (GR), deep resistivity (REID), self-potential (SP), and compressional acoustic (sonic) log (DT). When available, the density (DEN) and PhotoElectric Factor (PE) log data were considered.

The caliper log (CAL) serves as an indicator for the quality of the recordings of the other measured parameters. The experimental results highlight the benefits of selecting the data based on CAL log, benefits which in the context of ε-SVR are not limited only to increased accuracy but also influence the complexity of the model.

Table 4. The impact of normalization procedure on the prediction accuracy of the trained model (DT=*f*(GR, REID)) measured by MSE and MRE on the training and test sets

	Training set: T&T 1-10 (complete)			Test set: Whittenberg 3-29		
	MSE	MRE	Pearson Correlation index	MSE	MRE	Pearson Correlation index
Without normalization	66.4005	0.0548	0.8491	166.1475	0.0956	0.6611
With normalization	83.7345	0.0634	0.8039	126.7733	0.0951	0.7801

3.2.4. Data Normalization

Data normalization is a pre-processing step very popular in data analysis. One common use is to speed-up the convergence of the gradient descent algorithm that constitutes the basis of many machine learning tasks.

Data normalization is not mandatory in all regression algorithms, i.e. symbolic regression with genetic programming (Cranganu and Bautu, 2010) is not influenced by normalization. However, in ε-SVR, depending on the kernel used, data normalization may be crucial for the performance of the trained model.

All data sets are normalized as follows. For each feature in the training set the mean and the standard deviation are computed. These values are further used to scale each feature by subtracting from each observation the mean and dividing the result by the standard deviation. For all data sets involved in the experiments normalization is conducted using the mean and standard deviation computed on the training set. This normalization scheme guarantees mean 0 and variance 1 for the features in training set and comparable values for the other data sets.

Experimental results are very illustrative for the importance of normalization: we trained ε-SVR with radial kernel on the entire T&T 1–10 well consisting of 19,898 data values, to predict DT based on REID and GR parameters. To assess the quality of the trained regression function we used as test data values recorded in the Whittenberg 3–29 well (Figure 25).

We achieved a MSE value equal to 126.773 when normalization was used vs. a value of 166.148 when the data sets were not normalized. Also the number of support vectors varied significantly: 13,522 for the normalized version versus 19,341 when no normalization was performed. This difference in performance is mainly justified by the radial kernel which makes use of the Euclidean metric: due to different scales, one feature may dominate the others in distance calculations. Normalization improves not only the accuracy of predictions but also works towards better compression on the trained model. Table 4 illustrates the impact of normalization procedure on two types of errors (MSE and MRE) measured on the training and test data sets corresponding to two different wells.

All further experiments are performed on normalized data.

3.2.5. Training Set Selection

In supervised learning the output model (in our case the regression function) is created by learning patterns from the examples provided. Therefore, the training set should be carefully chosen in order to provide correct examples: noise should be removed, otherwise errors may be interpreted as existent patterns. At the same time the number of examples should be large enough to capture as many situations as possible.

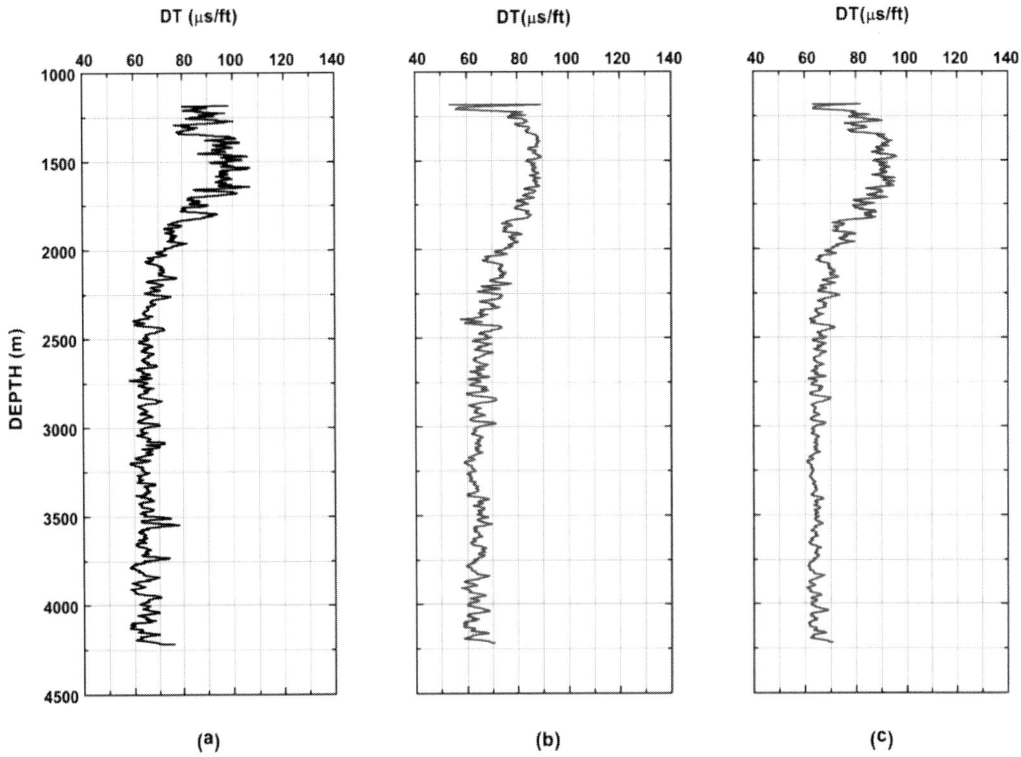

Figure 25. The effect of normalization on the performance of the model (DT= f(GR, REID)) measured on the test data Whittenberg 3-29. (a) the real DT values; (b) the predicted values corresponding to the model trained on non-normalized data; (c) the predicted values corresponding to the model trained on normalized data.

Following previous studies (Cranganu, 2005a, 2007), the T&T 1–10 well is used to build the training set. Based on the analysis of the CAL log, seven different depth intervals are chosen to be used by SVR. Best recorded data come from sequences of data where the diameter (described by CAL) is (almost) constant. Data that are suspicious (with respect to

CAL) are not included in the training set. Therefore, out of the approximately 20,000 data items available for the training well, the actual training set consists of 1,391 recordings in the following depth intervals: 1,570–1,600 m; 2,393–2,423 m, 2,713–2,743 m, 3,399–3,429 m, 3,581–3,612 m, 3,658–3,688 m, and 3,947–3,978 (Figure 26).

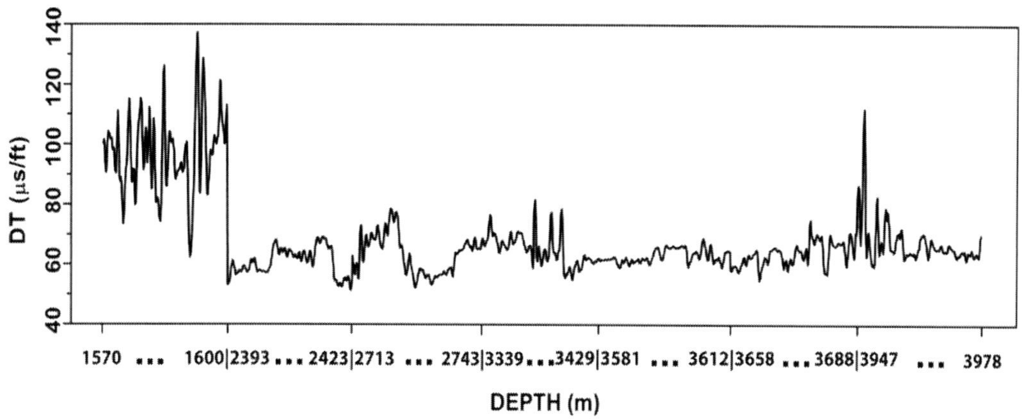

Figure 26. The reduced training set built based on the analysis of the CAL log from T&T 1–10 well.

Our experiments show that applying a filter based on the CAL parameter to clean the training data brings important improvements on the accuracy of predictions when the trained model is applied to derive predictions on new data sets. In a first scenario, when the regression model is trained on the entire data set T&T 1-10, the predictions on the Whittenberg 3-29 well achieve a MSE score of 126.773 (Table 4). In a second scenario, when the regression model is trained on the filtered/reduced T&T 1-10 data set, the predictions on the Whittenberg 3-29 well achieve a significantly lower MSE score of 100.005 (Table 5). The model trained on the reduced data set benefits also from reduced complexity consisting of only 771 support vectors compared to 13,522 support vectors for the model trained on the entire data set.

Table 5. Results obtained when the regression function (DT=*f*(GR, REID)) was trained on a reduced data set (T&T 1–10 reduced) built based on the analysis of the CAL log from the T&T 1–10 well. The MSE, MRE, and the correlation corresponding to the predictions of the model on: 1) the training set (T&T 1–10 reduced), 2) all available data from T&T 1–10 well (T&T 1–10 complete) which include the training data, and 3) Whittenberg well

	MSE	MRE	Pearson correlation index
Training set: T&T 1-10 reduced	35.713	0.044	0.903
Test set: T&T 1-10 complete	78.953	0.071	0.797
Validation set: Whittenberg 3-29	100.055	0.089	0.788

3.2.6. Parameter Selection

All experimental results shown previously were obtained with standard values for the SVR-specific parameters: C= 1, $\varepsilon = 0.1$, and radial kernel with $\sigma = 1$. But these values may not lead to the best model for our problem.

We performed a grid search over the three parameters. The model is trained on the T&T 1–10 well reduced as explained previously.

Two scenarios are tested in order to choose the best SVR-parameters:

1. *n*-fold cross validation, that involves only one data set - the training data set
2. a second data set is used for validation: the best parameters are chosen to be those that give the best accuracy when using the trained model for predictions on the validation set.

For cross-validation the T&T 1–10 reduced data set was split randomly into *n* groups of equal size: the model is trained on *n*-1 groups and the performance is assessed on the group left out. For one parameter setting there is a total of n trained models corresponding to all combinations of *n*-1 groups and a total of *n* MSE values corresponding to the performance of these models recorded on the groups left out. The performance of one of the parameters settings under test is thus computed as the average performance (average MSE) of the *n* trained models on the *n* groups combinations and the winning setting is chosen to be the one with the best average performance. A new regression function is trained on the T&T 1–10 reduced data using the best found parameters setting and then applied to other data sets playing the role of test data: Whittenberg 3-29, Barbara 1-20, Don-Elsie 1-20, Stubbs 2-35, Church-Pierce 1-3 – data sets which did not participate in the training phase. The results obtained for 5-fold and 10-fold cross-validation are presented in Table 6. In both cases, significant better prediction accuracy is obtained on the training set under these settings, compared to the use of standard values for the parameters of the method: an MSE value of 26.053 (Table 6) compared to a MSE value of 35.713 (Table 5). However, on the Whittenberg 3-29 test set the situation is opposite: the MSE values are greater than 150 as indicated in Table 6 when the parameters are optimized, while a MSE value of 100.055 was obtained when no parameter optimization was introduced in the analysis (Table 5). These results indicate that the best parameters settings found (and consequently the regression models) overfit the T&T 1–10 data set and fail to predict accurately other data sets. This phenomenon further indicates that the data in the training well and in the test well are heterogeneous.

The use of cross-validation to identify the parameters setting leading to the best regression model is prone to overfitting due to the use of a data set derived from only one training well (Figure 27).

The use of a distinct validation set is required under these circumstances. For a given parameters setting one regression model is trained on the T&T 1–10 reduced data set and its performance is assessed on the Whittenberg 3–29 data set (used here as a validation set). The best parameters setting and consequently the best regression model are chosen to be the one giving the best predictions on Whittenberg 3–29. For this data set a slight improvement is achieved over the case when standard values for the SVR parameters are used: a MSE value of 98.920 (Table 6, last row) over 100.055 (Table 5) on the Whittenberg 3-29 data set. The optimum parameters values corresponding to this result are: $C = 2$, $\varepsilon = 0.05$, and radial kernel with $\sigma = 1$. Because Whittenberg 3-29 plays the role of the validation set here, the regression model is further tested on the other data sets mentioned earlier for cross-validation. All the results obtained on the test sets and presented in Table 6 indicate that the regression model chosen with the aid of data coming from distinct well benefits from significantly improved generalization capability.

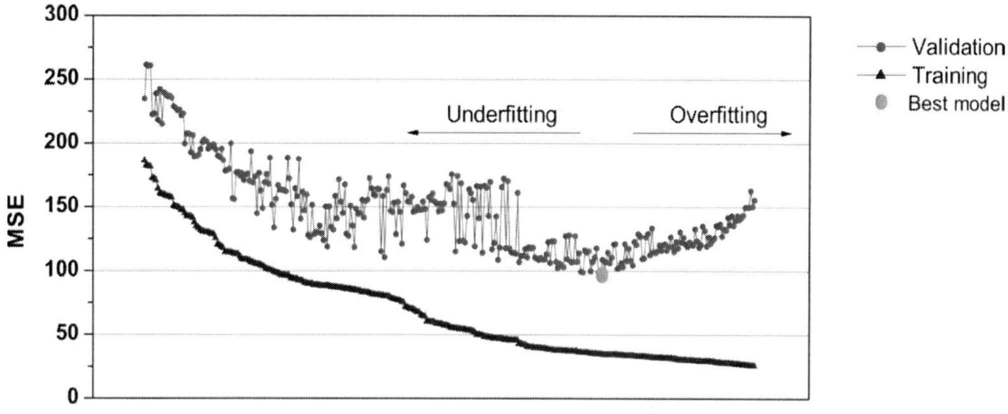

Figure 27. ε-SVR parameter selection using a validation set. MSE is plotted for models (DT = f(GR, REID)) trained on the reduced T&T 1-10 data set with various parameters configurations: the bottom curve corresponds to predictions on the training set and the top curve corresponds to predictions on the validation set Whittenberg 3-29.

Table 6. The prediction performance of the models (DT=*f*(GR, REID)) derived using different strategies for ε-SVR parameters selection: cross-validation (5-fold and 10-fold) involving only the training set (T&T 1–10 reduced) and validation involving Whittenberg 3-29 as validation set

		Training set T&T 1-10 reduced	Test set Whittenberg 3-29	Test set Barbara 1-20	Test set Don-Elsie 1-20	Test set Stubbs 2-35	Test set Church-Pierce 1-3
5-fold c-v	MSE	26.035	150.628	145.053	298.420	169.065	120.703
	cor	0.930	0.696	0.669	0.474	0.568	0.339
10-fold c-v	MSE	26.053	163.037	163.347	323.467	161.473	120.304
	cor	0.930	0.673	0.630	0.438	0.591	0.359
Using validation set	MSE	34.854	98.921	76.249	197.409	144.951	109.234
	cor	.906	0.797	0.824	0.662	0.608	0.308

Table 7. Prediction of DT based on various combinations of variables/predictors: the number of sup-port vectors for each model and the prediction performance measured with MSE, MRE and Pearson correlation index on the training set and validation set

Predictors	# support vectors	Training set: T&T 1-10 (reduced)			Validation set: Whittenberg 3-29		
		MSE	MRE	cor	MSE	MRE	cor
REID+GR+SP	190	39.108	0.058	0.899	133.833	0.105	0.700
GR+SP	1338	170.130	0.092	0.435	223.632	0.149	0.324
REID+SP	327	55.572	0.073	0.843	121.875	0.095	0.716
REID+GR	1033	34.854	0.042	0.906	98.921	0.084	0.797
SP	457	190.562	0.107	0.066	244.758	0.136	0.280
GR	1196	176.714	0.100	0.327	232.192	0.152	0.273
REID	207	38.2482	0.060	0.900	85.458	0.084	0.813

Figure 27 illustrates the overfitting and underfitting phenomena for SVR training with 350 different parameter settings taking values as follows: $C \in [0.01, 4]$, $\varepsilon \in [0.01, 1]$, $\sigma \in [0.1, 8]$. The training set used is T&T 1–10 reduced and the validation set used is Whittenberg 3-29. Bad performance of the trained model (high MSE) for both training and validation sets corresponds to underfitting while very good performance on the training data but poor performance on the validation data correspond to overfitting. The model resulting in the best performance on the validation data set is considered to be the best with regard to the capability of generalization to unseen data while obtaining fairly good prediction accuracy on both training and new data.

3.2.7. Predictors Selection

This experimental section aims at finding the best predictors for the sonic log.

3.2.7.1. Analysis of the Most Common Well Logging Parameters: GR, REID, SP

The most common well logging parameters recorded are GR, REID, and SP. The simplest measure indicating dependence between two numerical parameters is the Pearson correlation index (henceforth noted as *cor*). The correlation between DT and the above-mentioned variables on various data sets suggest that the most promising predictor is REID. No clear distinction can be made with regard to the other two parameters as the correlations recorded on different data sets disagree. However, the Pearson correlation index catches only linear relationships between variables. Therefore, we trained several □-SVR models consisting of various combinations of predictors (variables) and assess the prediction power of the variables by comparing the performance of these models.

Based on the previous results, we use the following experimental setup. The T&T 1-10 reduced data set is used for training and all variables are normalized. Parameter selection for ε-SVR with radial kernel is performed by a grid search using the Whittenberg 3-29 data set for validation.

Table 7 presents the performance of various combinations of predictors: the complexity of the models is given by the number of support vectors and the prediction accuracy is measured with *MSE*, *MRE* and Pearson correlation index.

The results obtained on the validation set indicate that the most suitable predictor for DT is REID, being able to obtain high accuracy when used alone. In conjunction with GR it behaves comparable well while adding SP as a third predictor decreases consistently the prediction accuracy (Table 7 and Figure 28).

To judge the benefits of using GR as a second predictor in conjunction with REID, Table 8 presents the prediction accuracy of the models trained above when applied on other data sets. It is again noticed that, when used alone, REID provides better predictions on four data sets (Barbara 1-20, Stubbs 2-35, and Church-Pierce 1-3). When the pair REID+GR is used as predictors, only Don-Elsie 1-20 data set shows better *MSE* and *cor* values.

Taking into account the complexity of the model (given by the number of support vectors) along with the *MSE* scores we may conclude that, with regard to the three predictors analyzed here, the best model is the one involving only one variable - REID. However, in practice it might be the case that the variable is prone to noise. In this regard it is advisable to use models consisting of several variables that, with their contribution in the regression function may bias the erroneous decision triggered by the error appeared in the input space of one single variable.

3.2.7.2. Analysis of Other Well Logging Parameters: DEN and PE

Variables like DEN and PE may be better predictors than GR, as indicated by the correlation values between them and DT on a fraction of data for which they are available. However, due to the lack of records regarding them, the experimental analysis is incomplete. Only two data sets are involved in the analysis of these predictors: the data set corresponding to the T&T 1-10 well and the data set corresponding to the Whittenberg 3-29 well. Both data sets are much reduced compared to previous experiments due to the elimination of the records having missing values for the two parameters under analysis: only 9,361 records are available for the T&T1-10 well and 3,894 for the Whittenberg 3-29 well.

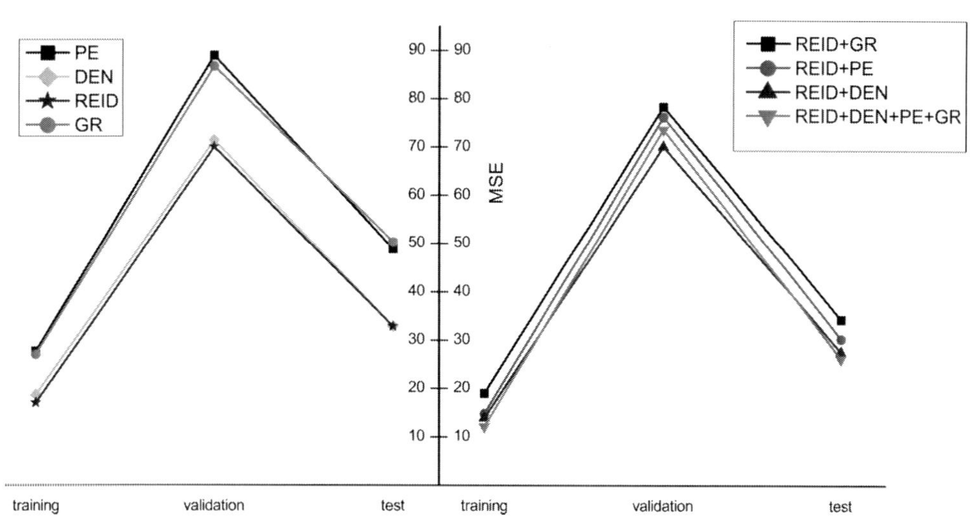

Figure 28. Comparative performance (computed MSE) of several models.

The following experimental setup is used in this part of the analysis.

The training set is built based on the T&T 1-10 well and reduced as in previous experiments based on the CAL parameter; from 1,391 records, those corresponding to missing values for DEN or PE parameters are eliminated resulting in a total of 801 records corresponding to the depth intervals 3,399-3,429, 3,581-3,612, 3,658-3,688 and 3,947-3,978.

As in previous experiments, the parameters of the method are optimized by a grid search. The validation set required in this step is Whittenberg 3-29 reduced to 3,894 records.

On these data sets a comparative experimental analysis is carried on involving all available well logging parameters. Various combinations of these predictors are used to train models to predict the DT parameter. The performance of these models on the training, validation and testing set is presented in Table 9 in a forward selection manner: to the current predictors set each new predictor is added in turn and the best combination is kept.

Figures 29 and 30 illustrate the predictions obtained with several models (REID, REID+GR, REID+DEN) on the validation set (Whittenberg 3-29) and on the test set (T&T 1-10 containing all 9,361 records). It is evident that the best predictions are achieved with the model using both REID and DEN as predictors. Compared to other models it is capable of predicting with higher accuracy the variations of the DT values. The other models tend to smooth the data and do not capture extreme variations.

We also stated in section 3.2.7.1 that models based on several predictors are preferred to models based on only one predictor when (inherently) noise is present in data. We validated experimentally the hypothesis by introducing normal noise in data for the predictor variables REID and DEN. For both variables we generated random numbers following a normal distribution with mean 0 and standard deviation ten times lower than the standard deviation of the predictor variables. Thus, we modified the Whittenberg data set used for validation as follows: REID = REID + (N(0, sd(REID)/10)) and DEN = DEN+N(0, sd(DEN)/10). The model based on REID alone achieved on the noisy Whittenberg a MSE value equal to 71.875. The model based on the joint used of REID and DEN achieves a MSE value of 70.149 when only the REID variable is noisy and a value of 70.259 when both are noisy. This indicates that the joint model is more robust in the presence of noise in variables.

Table 8. Results obtained on various test sets with the models trained previously on the T&T 1-10 reduced data set

	Test set: Barbara 1-20		Test set: Don-Elsie 1-20		Test set: Stubbs 2-35		Test set: Church-Pierce 1-3	
	MSE	cor	MSE	cor	MSE	cor	MSE	cor
REID+GR	76.249	0.824	197.409	0.662	144.952	0.608	109.234	0.308
REID	58.652	0.865	205.992	0.648	103.281	0.857	89.643	0.770

Because the joint use of REID and DEN as predictors does capture with high accuracy the variability of the DT parameter, we recommend it for future estimation of missing DT log.

3.2.8. Application of the Proposed Method

Once a regression model was trained and proved to obtain high prediction accuracy on unseen (test) data, it can be applied to predict unknown values on new data. At this stage we use some of the models built in the previous section to estimate the sonic log values DT in the application wells Smith 1–13 and Ledbetter 1–18 for which only the input parameters GR and REID are available.

Table 9. The performance of various combinations of predictors for predicting the sonic log DT. T&T1-10 is used for training and Whittenberg 3-29 is used for validation

Predictors	Training set: T&T 1-10 (reduced – 801 records)			Validation set: Whittenberg 3-29			Training set: T&T 1-10 (complete – 9,361 records)		
	MSE	MRE	cor	MSE	MRE	cor	MSE	MRE	cor
GR	27.068	0.045	0.245	86.777	0.087	0.203	50.345	0.070	0.090
PR	27.678	0.053	0.394	88.991	0.093	0.144	48.828	0.070	0.492
DEN	18.684	0.048	0.582	71.399	0.076	0.460	32.816	0.065	0.592
REID	17.058	0.037	0.631	69.944	0.070	0.482	32.850	0.053	0.601
REID+GR	18.997	0.0462	0.617	78.410	0.073	0.476	34.312	0.056	0.618
REID+PE	14.842	0.0331	0.699	76.202	0.065	0.406	30.360	0.051	0.622
REID+DEN	13.858	0.039	0.731	70.060	0.067	0.467	27.562	0.055	0.688
REID+DEN+GR	11.837	0.034	0.771	72.069	0.067	0.467	27.839	0.052	0.679
REID+DEN+PE	11.534	0.032	0.776	72.828	0.063	0.445	28.848	0.051	0.654
REID+DEN+PE+GR	12.126	0.037	0.770	73.606	0.064	0.456	26.238	0.051	0.700

The most two successful regression models trained in section 3.2.7.1 are used here for predictions: the one based only on the REID parameter and the one based on both REID and GR parameters. Figures 31 and 32 illustrate the DT values estimated in the application wells.

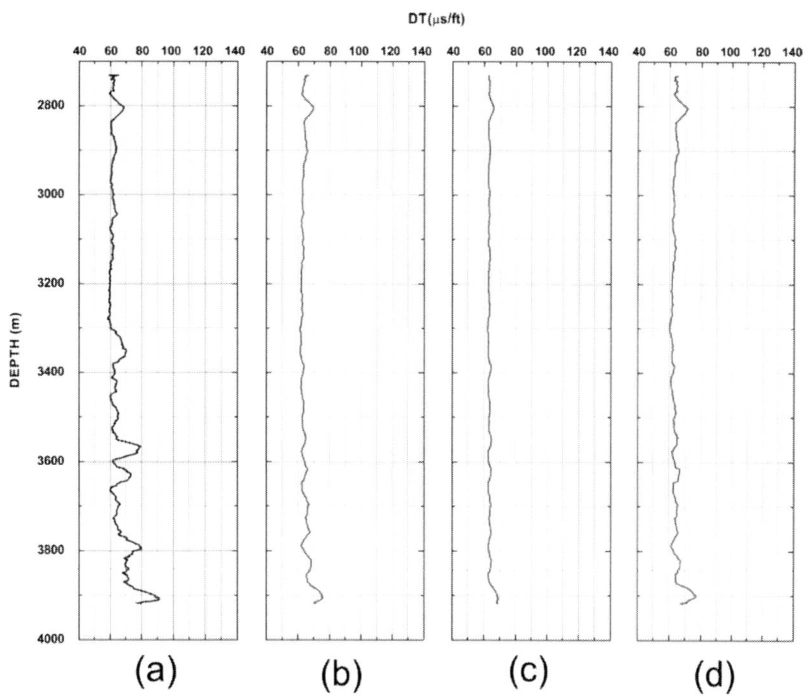

Figure 29. DT values for Whittenberg 3-29 data set: (a) the real values; (b) predictions based on REID; (c) predictions based on {REID and GR}; (d) predictions based on {REID and DEN}.

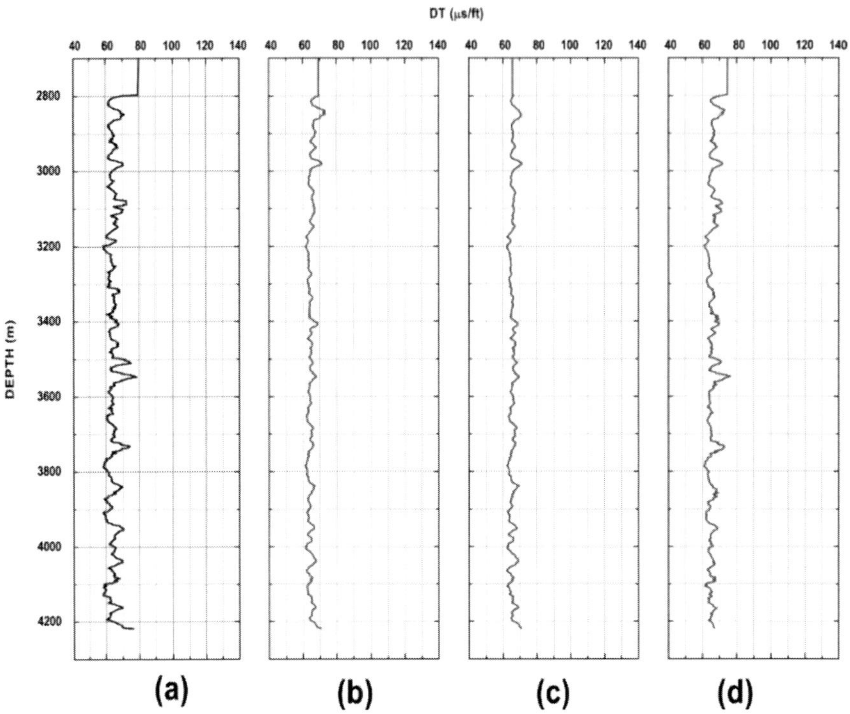

Figure 30. DT values for T&T1-10 data set; from left to right: a) the real values; b) predictions based on REID; c) predictions based on {REID and GR}; d) predictions based on {REID and DEN}.

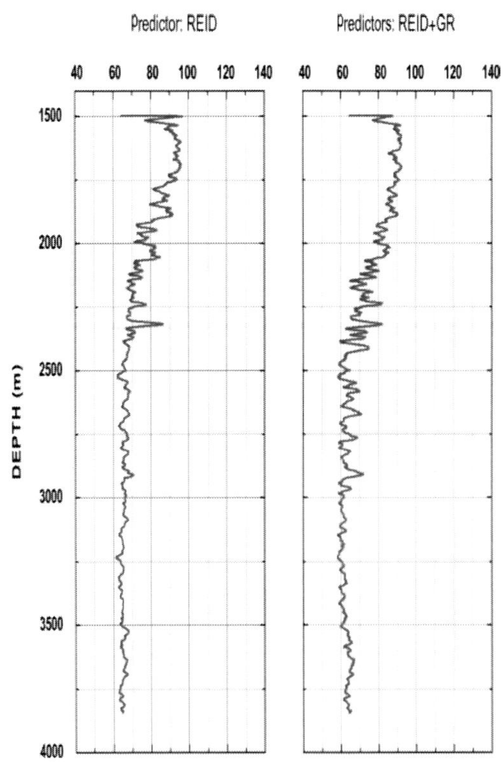

Figure 31. Predicted DT for application well Smith1–13.

3.3. Estimating the Presence of Overpressured Zones in the Anadarko Basin Based on Sonic Logs

It is well documented (e.g., Huffman and Bowers, 2002; Mitchell and Grauls, 1998; Surdam, 1997; Ortoleva, 1994) that many sedimentary basins throughout the world exhibit areas with abnormal pore-fluid pressures (higher or lower than normal or hydrostatic pressure). Predicting pore pressure and other related parameters (depth, extension, magnitude, etc.) in such areas is of vital importance for drilling operations and, in the same time, a challenging task. The compressional acoustic (sonic) log (DT) is often used as a predictor because it responds to changes in porosity or compaction produced by abnormal pore-fluid pressures (Cranganu, 2007; Cranganu and Bautu, 2010). Determining trends in acoustic transit time on sonic logs (DT) is a conventional method of predicting abnormal pressure in shales or sandstones interbedded with shales (e.g., Hottman and Johnson, 1965; Magara, 1976; Powley, 1982; Surdam et al., 1997; Hsu and Belaud, 1998). Because the velocity of a compressional wave in sedimentary rocks is dependent on the effective stress, an increase in transit time is interpreted as indicating a lower effective rock stress and, consequently, overpressuring. Hermanrud et al. (1998) found that the sonic and resistivity log-derived porosities are significantly higher in the overpressured than in normally pressured wells. Similarly, Poston and Berg (2002) state that sonic logs show an increase in interval transit times in overpressured shales. The calculation of abnormal pressure in pressure from the log data is based on the relationship between observed resistivity or acoustic values of shales and a trend line reflecting normal compaction (Heppard et al., 1998).

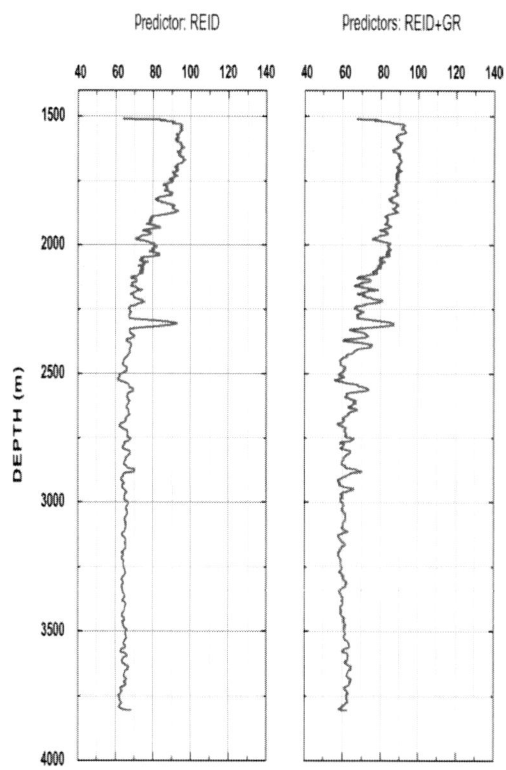

Figure 32. Predicted DT for application well Ledbetter 1–18.

These considerations have led us to predict the presence of overpressured zones in the training well T&T 1-10, between ~1,250 m – 1,950 m), in the verification well Whittenberg 3-29 between ~1,350 m – 2,000 m, as well as in the two application wells chosen for this article: Smith 1-13 well, between ~ 1,500m – 2,000 m and Ledbetter 1-18 well, between ~1,500m – 1,800 m (Figure 33).

These depth intervals are mostly sandy shales with intercalations of shaly sandstones (Cranganu, 2005b). The presence of such overpressured zones (compartments) within the Anadarko Basin is well documented (Al-Shaieb et al., 1990, 1997a, 1997b; Cranganu, 2005a, 2005b, 2007). Our estimated overpressured intervals are consistent with previous measured overpressured intervals, as detailed by Al-Shaieb et al., 1990, 1997a, 1997b.

3.4. Conclusion about Using SVR

We presented support vector regression as a tool for predicting the sonic log in gas/oil wells based on other available logs. Several experimental scenarios are proposed in order to highlight and formulate a set of good practices to be followed when SVR is to be used in the analysis of log data: data normalization, training set selection, and optimization of the SVR-parameters. All are important steps deciding the prediction power, the generalization capability and the complexity of the derived regression model. Extensive experiments are carried on to investigate the prediction power of several geophysical parameters and of their joint use for a specific goal behind this study: the detection of overpressured regions in gas/oil wells.

For the particular data sets we used, we showed that the most suitable predictor for the sonic log DT is the deep resistivity log REID, which is able to yield high accuracy when used alone. Adding other predictors, like the density log (DEN) and the gamma ray GR log, produces comparable results, but adding SP as predictor decreases consistently the prediction accuracy.

Considering both the complexity of the model built based upon the number of support vectors and the accuracy of predictions evaluated with MSE, we may conclude that the best SVR model is the one with a single predictor, REID. However, in practice, it is advisable to use models consisting of several variables because they are more reliable. When using only one predictor, any erroneous values on that predictor result in erroneous predictions. In the presence of other predictors, the error in one predictor may be counterbalanced by correct "votes" from the others. We found out that the joint use of REID and density DEN log captures with high accuracy the variability of DT and, therefore, is advisable to further use this combination for estimating DT values when both predictors are available.

We finally used the simulated DT values in four wells drilled in the Anadarko Basin, Oklahoma, to infer the presence and to estimate the extent of areas with overpressured pore fluids. The results obtained are in very good agreement with previous published works and demonstrate the capability of DT log to be used in such endeavors.

The method presented here is not limited to modeling DT values only. It can be extended, with appropriate modifications of the algorithm, in any area of well logging studies, where missing log values are needed. Thus, we offer a blueprint for future similar applications.

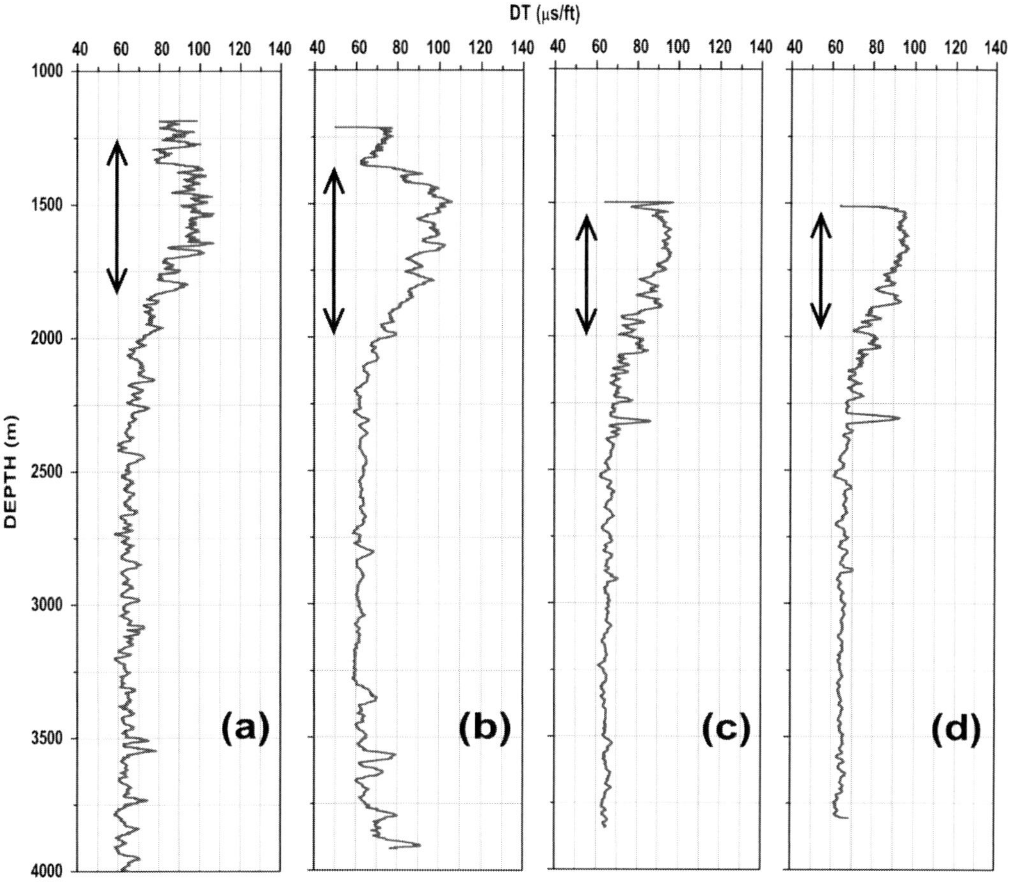

Figure 33. Estimated overpressured zone using computed DT values in the T&T 1-10 training well (a), Whittenberg 3-29 verification well (b), application well Ledbetter 1-18 (c), and application well Smith 1-13 (d). For the application wells, DT values computed using REID as predictor were used. The extent and location of overpressured zones is indicated by arrows.

REFERENCES

Accarain, P., and Desbrandes, R., 1993, Neuro-computing helps pore pressure determination, 1993, *Petroleum Eng. Int.,* February 1993, 39-42.

Al-Anazi, A. F., and Gates, I. D., 2010a, Support vector regression for porosity prediction in a heterogeneous reservoir: A comparative study, *Computers & Geosciences*, 36, 1494 – 1503.

Al-Anazi, A. F., and Gates, I. D., 2010b, Support-vector regression for permeability prediction in a heterogeneous reservoir: A comparative study, *SPE paper* 126339.

Ali, M., and A. Chawathe, 2000, Using artificial intelligence to predict permeability from petrographic data, *Computers and Geosciences*, 26, 915-925.

Al-Shaieb, Z., J. O. Puckette, P. B. Ely, and V. Tigert, 1990, Pressure Compartments and Seals in the Anadarko Basin, *In* Johnson, K.S., and Cardott, B.J. (eds.), Source Rocks in the Mid-continent, 1990 Symposium: *Oklahoma Geological Survey Circular* 93, 210-228.

Al-Shaieb, Z., J. O. Puckette, A. A. Abdalla, and P. B. Ely, 1997a, Megacompartment complex in the Anadarko Basin–a completely sealed overpressured phenomenon, *In* Gas Research Institute Workshop, Compartmentalization of Overpressured Reservoirs in the Anadarko Basin–Implications for Exploration and Development: *Gas Research Institute*, Oklahoma City, 1-14.

Al-Shaieb, Z., J. O. Puckette, A. A. Abdalla, P. B. Ely, 1997b, Three levels of compartmentation within the overpressured interval of the Anadarko Basin, *In* Gas Research Institute Workshop: Compartmentalization of Overpressured Reservoirs in the Anadarko Basin–Implications for Exploration and Development: *Gas Research Institute*, Oklahoma City, 15-29.

Arbogast, J. S., M. L. Buttler, M. H. Franklin, and K. A. Thompson, 2000, Enhancement of "limited" log suites using neural networks, *In* Johnson, K. S. (ed.), Pennsylvanian and Permian geology and petroleum in the southern Midcontinent, 1998 symposium: *Oklahoma Geological Survey Circular* 104, 185 – 195.

Aristodemou, E., C. Pain, C. de Oliveira, T. Goddard, and C. Harris, 2005, Inversion of nuclear well-logging data using neural networks, Geophysical Prospecting, 53, 103-120.

Asquith, G., and Krygowski, D., 2004, Basic well log analysis, *AAPG Methods in Exploration Series* 16, 244 p.

Baldwin, J. L., R. M. Bateman, and C. L. Wheatley, 1990, Application of a neural network to the problem of mineral identification from well logs, *The Log Analyst*, 3, 279-293.

Benaouda, B., G. Wadge, R. B. Whitmarhs, R. G. Torhwell, and C. MacLeod, 1999, Inferring the lithology of borehole rocks by applying neural network classifiers to downhole logs – an example from the Ocean Drilling Program, *Geophysical Journal International*, 136, 477-491.

Brameier, M., and Banzhaf, W., 2001, A Comparison of Linear Genetic Programming and Neural Networks in Medical Data Mining, *IEEE Transactions on Evolutionary Computation*, 5, 17–26.

Brameier, M., and Banzhaf, W., 2007, Linear Genetic Programming, Springer, 316 p.

Chang, C.-C., and Chih-Jen Lin, C.-J., 2011. LIBSVM: a library for support vector machines, *ACM Transactions on Intelligent Systems and Technology*, 2, 27:1--27:27.

Cortes C., and Vapnik, V., 1995, Support vector networks, Machine Learning, 20, 273–297.

Cranganu, C., 2004, Capillary sealing in the Anadarko Basin, Oklahoma, *Northeastern Geology and Environmental Sciences*, 26, 35 – 42.

Cranganu, C, 2005a, Using artificial neural networks to predict abnormal pressures in the Anadarko Basin, Oklahoma, *Journal of Balkan Geophysical Society*, 8, 1, 343 – 348.

Cranganu, C. 2005b, Looking for gas-bearing layers in the Anadarko Basin, Oklahoma, *Oklahoma Geology Notes*, 65, 3, 72-77.

Cranganu, C., 2006, Artificial Neural Networks May Help Predicting Abnormal Pressures in the Anadarko Basin, *AAPG 2006 Annual Convention, Abstract Volume*, 22.

Cranganu, C., 2007, Using Artificial Neural Networks to Predict the Presence of Overpressured Zones in the Anadarko Basin, Oklahoma, *Pure and Applied Geophysics*, 164, 2067 – 2081.

Cranganu, C., and Bautu, E., 2010, Using Gene Expression Programming to estimate sonic log distributions based on the natural gamma ray and deep resistivity logs: A case study from the Anadarko Basin, Oklahoma, *J. Pet. Sci. and Eng.*, 70, 243–255.

Cranganu, C., and Breaban, M., 2013, Using support vector regression to estimate sonic log distributions: A case study from the Anadarko Basin, Oklahoma, *J. Pet. Sci. and Eng.*, 103, 1 – 13, http://dx.doi.org/10.1016/j.petrol.2013.02.011

Dimitriadou, E., Hornik, K., Leisch, F., Meyer, D., and Weingessel, A., 2012, *Misc Functions of the Department of Statistics* (e1071), TU Wien, last accessed on line on March 26, 2012 at http://cran.r-project.org/web/packages/e1071/e1071.pdf

Drucker, H., Burges, C. J. C., Kaufman, L., Smola, A., and Vapnik, V., 1996. Support Vector Regression Machines. Advances in Neural Information Processing Systems 9, NIPS 1996, 155–161, MIT Press.

Ellis, D. V., and Singer, J. M., 2007, *Well Logging for earth Scientists*, Springer, The Netherlands, 692 p.

Ferreira, C., 2001, Gene Expression Programming: A New Adaptive Algorithm for Solving Problems, *Complex Systems*, 13, 2, 87 – 129.

Ferreira, C., 2002, Genetic Representation and Genetic Neutrality in Gene Expression Programming, *Advances in Complex Systems,* v. 5, issue 4, 389–408.

Ferreira, C., 2006, Gene Expression Programming: Mathematical Modeling by an Artificial Intelligence, 2nd ed., Springer, 478 p.

Fletcher, R., 1989, *Practical Methods of Optimization*, John Wiley and Sons, New York.

Friedman, J. H., 1991, Multivariate Adaptive Regression Splines, *Annals of Statistics*, 19, 1 - 67.

Futuyma, D. J., 2006, Evolutionary Biology, 3rd ed., *Sinauer Associates*, Sunderland, MA, 763 p.

Hadamard, J., 1902, Sur les problèmes aux dérivées partielles et leur signification physique. Princeton University Bulletin, 49 – 52.

Haykin, S., 1999, Neural networks: a comprehensive foundation, 2nd ed., Prentice-Hall. Inc., Upper Saddle River, New Jersey, 842 p.

Hearst, J. R., Nelson, P. H., and Paillet, F. L., 2000, Well Logging for Physical Properties – A Handbook for Geophysicists, Geologists, and Engineers, John Wiley and Sons, Ltd., New York, 483 p.

Heppard, P. D., Cander, H. S., Eggertson, E. B., 1998, Abnormal pressure and the occurrence of hydrocarbons in Offshore Eastern Trinidad, West Indies, *In* Law, B. E., G. F. Ulmishek, and V. I. Slavin (Editors), *Abnormal pressures in hydrocarbon environments.* AAPG Memoir 70, 215-246.

Hermanrud, C., L. Wensaas, G. M. G. Teige, E. Vik, H. M. Nordgård Bolås, and S. Hansen, 1998, Shale porosities from well logs in Haltenbanken (offshore mid-Norway) show no influence of overpressuring, *In* Law, B. E., G. F. Ulmishek, and V. I. Slavin, eds., Abnormal pressures in hydrocarbon environments: *AAPG Memoir* 70, 65-87.

Hottman, C. E., and R. K. Johnson, 1965, Estimation of formation pressures from log-derived shales properties: *Journal of Petroleum Technology*, v. 17, 717 – 772.

Hsu, K., and D. Belaud, 1998, Integration application of real-time LWD sonic for overpressure detection and seismic, B*ull. Centre Rech. Elf Explor. Prod., Mém*. 22, 221 – 226.

Huang, Z., J. Shimed, M. Williamson, and J. Katsube, 1996, Permeability prediction with artificial neural network modeling in the Venture gas field, offshore eastern Canada, *Geophysics*, 61, 422-436.

Huffman, A. R., and Bowers, G. L. (Eds.), 2002, Pressure regimes in sedimentary basins, *AAPG Memoir* 76, 238 pp.

Kadkhodaie-Illkhchi, A., and A. Amini, 2009, A fuzzy logic approach to estimating hydraulic flow units from well log data: A case study from the Ahwaz oilfield, south Iran, *Journal of Petroleum Geology,* 32, 67 – 78.

Karush, W., 1939, Minima of functions of several variables with inequalities as side constraints. *Master's thesis*, Dept. of Mathematics, Univ. of Chicago.

Keith, M. J., and Martin, M. C., 1994, Genetic Programming in C++: Implementation Issues, *In* K. E. Kinnear, Jr., (Ed.) Advances in Genetic Programming, *MIT Press,* 285 – 310.

Kirshenbaum, E., and Suermondt, H. J., 2004, Using Genetic Programming to Obtain a Closed-Form Approximation to a Recursive Function, *In* Kalyanmoy, D. et al., *Genetic and Evolutionary Computation Conference,* Seattle, WA, USA, June 26-30, 2004, Proceedings, Part II. Lecture Notes in Computer Science 3103, Springer, 543-556.

Koza, J. R., 1992, Genetic Programming: On The Programming Of Computers By Means Of Natural Selection, MIT Press, Cambridge, MA, 835 p.

Kuhn, H. W., and Tucker, A. W., 1951, Nonlinear programming, *In* Proc. 2nd Berkeley Symposium on Mathematical Statistics and Probabilistics, 481–492, Berkeley, University of California Press.

Ligtenberg, J. H., and A. G. Wansink, 2003, Neural network prediction of permeability in the El Garia formation, Ashtart oilfield, offshore Tunisia, *In* M. Nikravesh, F. Aminzadeh, and L. A. Zadeh (eds), *Soft computing and intelligent data analysis in oil exploration,* Elsevier, 397 – 421.

Luhti, S. M., and I. D. Bryant, 1997, Well-log correlation using a back-propagation neural network, *Mathematical Geology,* 29, 3, 413- 425.

Magara, K., 1976, Thickness of removed sedimentary rocks, paleopressure, and paleotemperature, southwestern part of Western Canada Basin, *AAPG Bulletin,* 60, 554 – 565.

Mitchell, A. Grauls, D. (Eds.), 1998, Overpressures in petroleum exploration, *Bull. Centre Rech. Elf Explor. Prod., Mem.* 22, 248 pp.

Nikravesh, M., and Aminzadeh, F., 2003, Soft computing for intelligent reservoir characterization and modeling, *In* M. Nikravesh, F. Aminzadeh (Eds.), Developments in Petroleum Science 51, *Elsevier Science B.* B., 3 – 32.

Ortoleva, O. J., (Ed.), 1994, Basin compartments and seals, *AAPG Memoir* 61, 477 pp.

Poston, S. W., and R. R. Berg, 2002, Overpressured gas reservoirs, *SPE*, Richardson, TX, 138 p.

Powley, D. E., 1982, Pressures, normal and abnormal: AAPG Advanced Exploration Schools Unpublished Lecture Notes, 38 p.

Reed, R. D., and R. J. Marks, 1999, Neural smithing, supervised learning in feedforward artificial networks, *The MIT Press*, var. no p.

Rogers, S. J., H. C. Chen, D. C. Kopaska-Merekel, and J. H. Fang, 1995, Predicting permeability from porosity using artificial neural networks, *AAPG Bulletin*, 79, 1786-1797.

Romero, C. E., and Carter, J. N., 2001, Using genetic algorithms for reservoir characterization, *Journal of Petroleum Science and Engineering,* v. 31,113–123.

Romero, C. E., and Carter, J. N., 2003, Using genetic algorithms for reservoir characterization, *In* M. Nikravesh and F. Aminzadeh (Eds.), *Developments in Petroleum Science*, 51, Elsevier Science B.B., 323 – 363.

Saemi, M., Ahmadi, M., and Varjani, A. Y., 2007, Design of neural networks using genetic algorithm for the permeability estimation of the reservoir, *Journal of Petroleum Science and Engineering*, 59, 97–105.

Saggaf, M. M., and E. L. Nebrija, 2000, Estimation of lithologies and depositional facies from wire-line logs, *AAPG Bulletin,* 84, 1633-1646.

Saggaf, M. M., and E. L. Nebrija, 2003, Estimation of missing logs by regularized neural networks, *AAPG Bulletin,* 87, 1377-1389.

Schuelke, J. S., J. A. Quirein, J. F. Sarg, D. A. Altany, and J. H. Hunt, 1997, Reservoir architecture and porosity distribution, Pegasus field, west Texas – an integrated sequence stratigraphic-seismic attribute study using neural networks (abs.), 67th Annual International Meeting Society Exploration Geophysicists, Expanded Abstracts, 668- 671.

Surdam, R, C. (Ed.), 1997. Seals, traps, and the petroleum system, *AAPG Memoir* 67, 317 pp.

Surdam, R. C., Z. S. Jiao, H. P. Heasler, 1997, Anomalously pressured gas compartments in Cretaceous rocks of the Laramide Basins of Wyoming: A new class of hydrocarbon accumulation, *In* Surdam, R. C. (ed.), 1997, Seals, traps, and the petroleum system, *AAPG Memoir* 67, 199 – 222.

Vapnik, V., Golowich, S., and Smola, A., 1997, Support vector method for function approximation, regression estimation, and signal processing, *In* M. C. Mozer, M.I. Jordan, and T. Petsche, Eds., *Advances in Neural Information Processing Systems 9*, 281–287, Cambridge, MA, 1997. MIT Press.

Velez-Langs, O., 2005, Genetic algorithms in oil industry: An overview, *Journal of Petroleum Science and Engineering,* 47, 15– 22.

Vladislavleva, E., 2008, *Model-based Problem Solving through Symbolic Regression via Pareto Genetic Programming*, PhD thesis, Tilburg University, Tilburg, the Netherlands, 288 p.

Wong, P. M., M. Jang, S. Chao, and T. D. Gedeon, 2000, Multiple permeability predictions using an observational learning algorithm, *Computers and Geosciences*, 20, 907-913.

Yu, T., and David Wilkinson, D., 2008, A Co-Evolutionary Fuzzy System for Reservoir Well Logs Interpretation, *In* Yu, T., Davis, D., Baydar, C., and Rajkumar, R. (Eds.), *Evolutionary Computation in Practice*, vol. 88, 199-218, Springer.

Yu, T., Wilkinson, D., and Castellini, A., 2008, Constructing Reservoir Flow Simulator Proxies Using Genetic Programming for History Matching and Production Forecast Uncertainty Analysis, *Journal of Artificial Evolution and Applications,* vol. 2008, Article ID 263108, doi:10.1155/2008/263108.

Zellou, A. M., and A. Ouenes, 2003, Integrated fractured reservoir characterization using neural networks and fuzzy logic: Three case studies, in M. Nikravesh, F. Aminzadeh, and L. A. Zadeh (eds), *Soft computing and intelligent data analysis in oil exploration*, Elsevier, 583 – 604.

In: Natural Gas and Petroleum
Editor: Constantin Cranganu
ISBN: 978-1-62808-535-8

Chapter 3

RESERVOIR AND NON-RESERVOIR ROCKS FROM THE ANADARKO BASIN, OKLAHOMA: A PETROPHYSICAL OUTLINE

***Constantin Cranganu*[1,*] *and Maria A. Villa*[2]**
[1]City University of New York, Brooklyn College, NY, US
[2]Oxy-Petroleum, Bogotá, Colombia

ABSTRACT

Using mercury intrusion porosimetry technique we analyzed 91 core samples taken from the reservoir and non-reservoir rocks of the Anadarko Basin, Oklahoma. The samples are sandstones (20), shales (24), interbedded shales and siltstones (11), siltstones (17), grainstones (9) and lime mudstones (10), ranging in age from Devonian to Permian and in depth from 404 m to 4,975 m. The objectives of this study are (1) to develop a classification of reservoir vs. non-reservoir rocks in the Anadarko Basin based on their lithology and petrophysical characteristics; (2) to establish the relationships between various petrophysical properties, lithology, and reservoir quality; 3) to determine how these relationships are affected by sorting of the pore-throat size distribution, and (4) to build a petrophysical database for reservoir and non-reservoir rocks of the Anadarko Basin. The results of this study will help resolve some of the key challenges faced in both exploration for and production of hydrocarbon: using only those petrophysical characteristics that can clearly distinguish between reservoir and non-reservoir rocks, identifying and characterizing petrophysical flow units, indicating the number of significant fluid invasions, understanding the relationships between petrophysical properties and lithology, or inferring the reservoir quality and performance.

The petrophysical parameters used in this study are: median pore-throat radius (MPR), average pore-throat radius (APR), apparent porosity (Φ), air permeability (Kair), pore-throat radius at 35% mercury saturation (R35), maximum threshold-entry radius (MTER), pore-throat size distribution (PTD), port type (PT), and pore-throat sorting (PTS). The quality and performance of reservoir rocks was estimated using a permeability/porosity/port type plot. Influence of lithology on petrophysical parameters

* Corresponding author: 2900 Bedford Avenue, Brooklyn, NY, USA, Email: cranganu@brooklyn.cuny.edu.

was investigated. Finally, statistical analysis was used to assess correlations between various petrophysical parameters.

APR, MTER and Φ values are not well suited to discriminate between reservoir and non reservoir rocks (their value intervals mostly overlap). We assume that this situation is due to lithological heterogeneity of samples.

MPR, R35, and Kair have distinct value intervals for reservoir and non-reservoir rocks. Our interpretation is that the main control is exerted by port types: mostly mega and macro port types for reservoir rocks and mostly micro and nano port types for non-reservoir rocks.

Reservoir rocks display a majority (36 out of 50 samples) of bimodal pore-throat distributions, while all non-reservoir rocks (41 samples) have unimodal distribution. PT and PTS are considered responsible for this situation.

Another significant distinction between reservoir and non-reservoir rocks is provided by PTS: it is dominated by PS (poorly sorting) and MS (medium sorting) for reservoir rocks and by WS (well sorting) and MS for non-reservoir rocks. The link between PTS and lithology is clearly illustrated. Most samples that have WS PTS are sandstones (44%); most samples that exhibit MS PTS are siltstones (32%), whereas most samples with PS PTS are shales (33%). The distributions are significantly different ($\chi^2 = 23.606$, DF = 10, p-value = 0.009).

Two relationships (Φ vs. APR and Φ vs. MTER, both for reservoir rocks) have statistical significant correlation with p-value <0.05. None of statistical correlations has a statistical significance for non-reservoir rocks. We infer that lithological heterogeneity, depositional environments, lateral and vertical change in the formation's mineral composition and/or different port types may be responsible for such situations.

Although modest, the data collected during this research represent the beginning of a petrophysical database for reservoir and non-reservoir rocks of the Anadarko Basin, OK.

In this study we illustrated the various links between petrophysical characteristics and distribution of reservoir and non-reservoir rocks in the Anadarko Basin, Oklahoma.

This study produced (1) a classification of reservoir vs. non-reservoir rocks based upon their lithology and various petrophysical parameters; (2) identified criteria for classifying reservoir and non-reservoir rocks based on capillary-pressure curves; and (3) established statistical relationships between lithology and petrophysical characteristics of the reservoir and non-reservoir rocks in the Anadarko Basin, OK.

Keywords: Reservoir rocks, non-reservoir rocks, petrophysical parameters, Anadarko Basin

INTRODUCTION

The Anadarko Basin is the deepest sedimentary basin on the North American craton. Total sediment thickness on the deepest section exceeds 40,000 ft (12 km) and consists of sandstones, limestones, and shales of Cambrian to Permian age (Johnson, 1989). The basin covers almost the entire western part of Oklahoma and extends into the northeastern part of the Texas Pan-handle (Figure 1).

The Anadarko Basin is one of the most prolific oil and natural gas producers in the North American continent. Cumulative production by the end of 2004 was almost 12.7 billion barrels of oil and more than 12 trillion cubic feet of gas (Boyd, 2005). The Anadarko Basin has produced more oil and gas than any other petroleum province in the state of Oklahoma: the Arkoma Basin, the Cherokee Platform, the Ouachita Foldbelt, or the Southern Oklahoma Foldbelt.

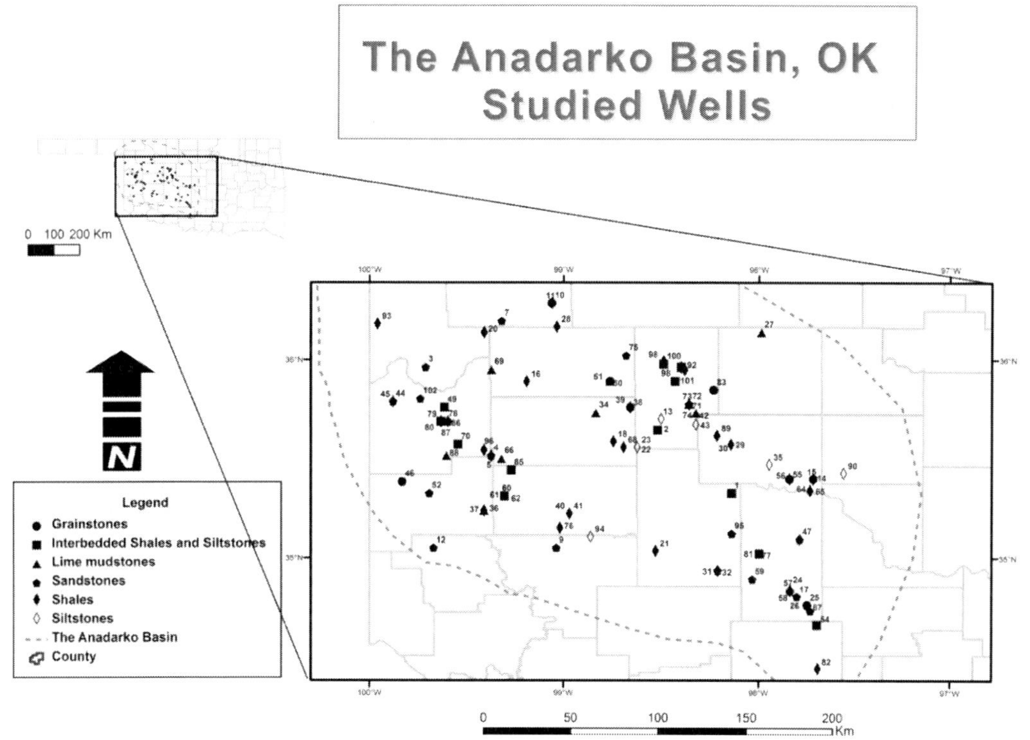

Figure 1. The Anadarko Basin, OK. Sample locations and lithological types. The numbers are the same as ID numbers in Table 1 (this notice applies to all figures).

Nowadays the main production of the Anadarko Basin is natural gas. The majority of the basin's gas reserves occur in the basin-scale compartments termed the mega-compartment complex (MCC) (Al-Shaieb et al., 1992, 1994a, and 1994b) and in the deep basin below the MCC. The upper part of the basin also contains substantial amounts of oil and natural gas. In 1980, the ultimate recovery of natural gas from the Anadarko Basin was estimated at more than 110 TCF (trillion cubic feet) (Hill and Clark, 1980), an amount approximately equivalent to four times the year 2000 production of natural gas from all of North America. Several large (> 1 TCF production through January 2001) gas fields are located in the Oklahoma part of the Anadarko Basin. According to Lay (2001), these fields - Elk City, Putnam, Carpenter, Watonga-Chickasha, as well as portions of the Sooner Trend and Golden Trend fields - have produced about 9 TCF of gas since their discovery and exploitation. The Anadarko Basin is also believed to be the source of gas found in the nearby Panhandle-Hugoton field, the largest gas field in North America (Lay, 2001). Clearly, the Anadarko Basin still holds large gas reserves that need to be discovered and exploited.

The outlook for the Anadarko Basin's oil and gas production depends on many factors, such as price, drilling activity, consumption, regulation, energy security, etc. Petrophysical characteristics and facies distribution of reservoir and non-reservoir rocks of the Anadarko Basin play an important role in determining the quality of existing and new reservoirs in area, thus enabling an enhanced oil and gas production. This kind of study requires recognition of the heterogeneity of reservoir and non-reservoir characteristics and delineation of petrophysical facies. The physical parameters and the lateral variability of reservoir vs. non-reservoir quality are directly linked to textures and pore-system dimensions.

Pore systems consist of relatively large voids, or pores, distributed among smaller passages called pore throats. Pore-throat sizes and their distribution in reservoir and non-reservoir rocks can be estimated by capillary-pressure curves derived from mercury intrusion porosimetry (MIP). The MIP data are obtained by forcing mercury at pressures up to 413,685 kPa (60,000 psi) into small voids and pore throats within the rock sample. Pore throats control access to larger voids (pores) because greater pressures are required to force mercury, or other nonwetting fluid, into smaller spaces (Purcell, 1949; Keighin, 1997). Thus pores are bottle-necks in the system, and it is necessary to exceed their critical capillary pressure in order to inject mercury into pores. Mercury injection pressure is increased in a stepwise manner and time for equilibration between pressure increments is allowed. The step pressure is plotted against mercury saturation (Figures 2, 3, and 4).

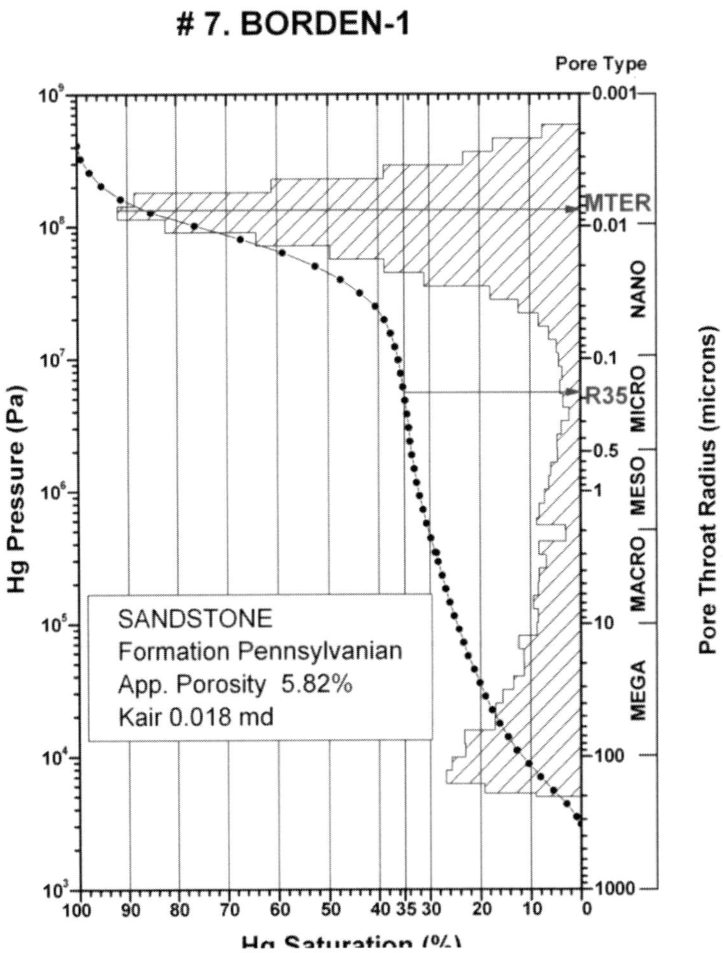

Figure 2. Capillary-pressure curve and petrophysical parameters derived from measurements. Sample ID#7, Borden-1 well. Apparent porosity: 5.82%; air permeability: 0.018 md, R35 (pore-throat radius at 35% mercury saturation): 0.11 μm; MTER (Maximum Threshold-Entry Radius): 0.0069 μm ; PT (Port Type): Micro; PTS (Pore-throat sorting): MS (medium sorting); PTD (Pore-throat size distribution): unimodal; Reservoir quality: window 3 (non-reservoir rock).

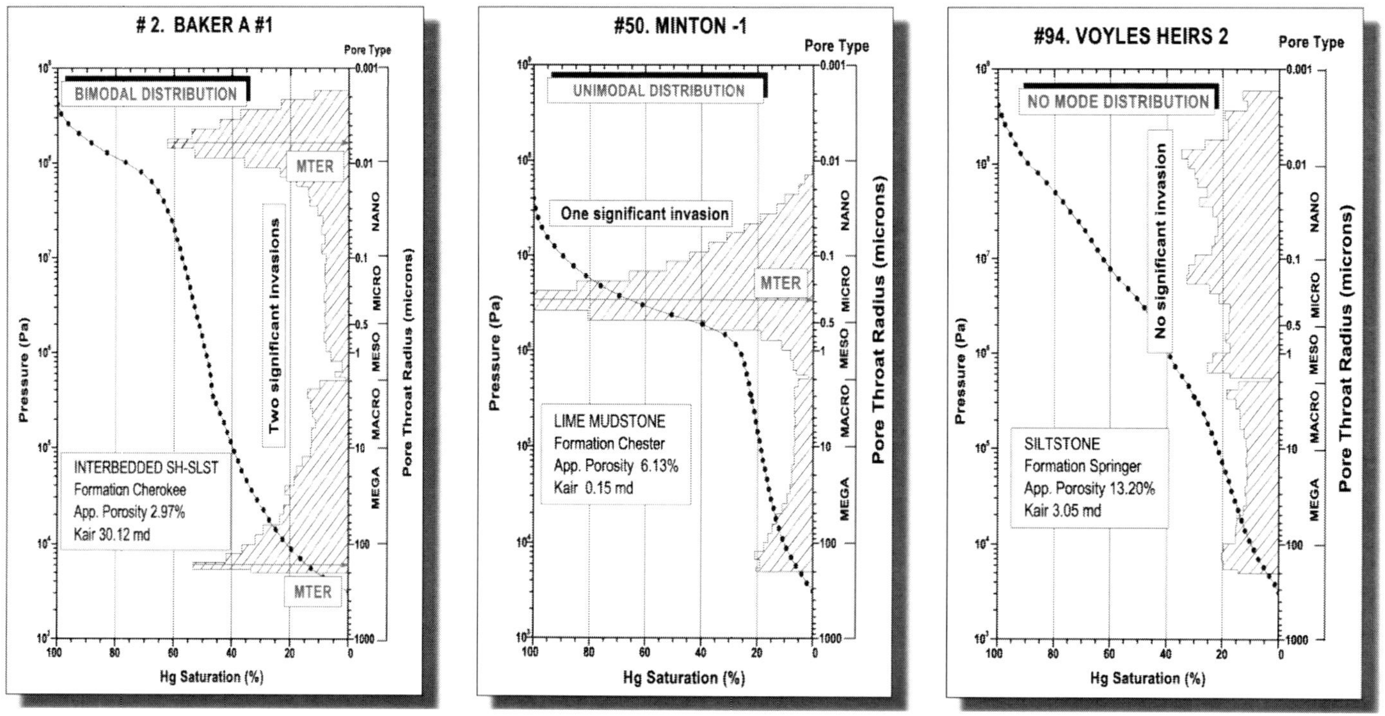

Figure 3. Pore-throat size distribution (PTD): unimodal (a single significant fluid invasion), bimodal (two significant fluid invasions, one at a higher pore-throat size and another at lower pore-throat size), and no mode (no significant fluid invasion is present).

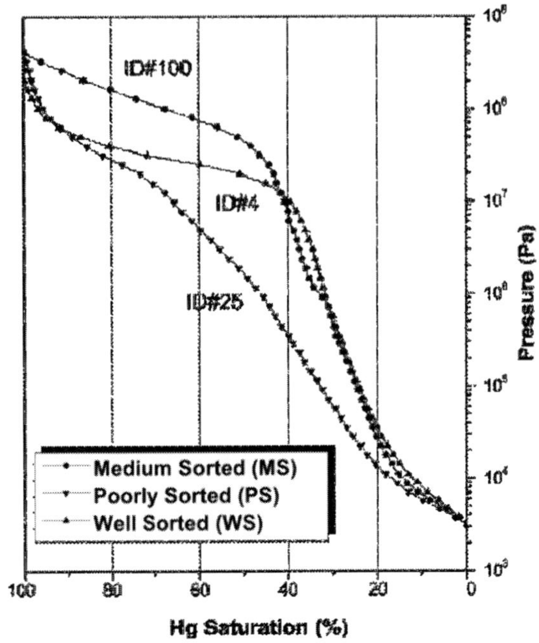

Figure 4. Pore-throat sorting (PTS): poorly sorted (PS), medium sorted (MS), and well sorted (WS).

The relationship between applied pressure P and the minimum size pore D into which mercury will be forced to enter is based on Washburn equation (Washburn, 1921):

$$D = -4\gamma \cos\theta / P \qquad (1)$$

where γ is the interfacial tension of mercury-air system (0.485 N/m), and θ is the air-mercury-solid contact angle (140°).

Petrophysical facies are laterally and vertically continuous units that have similar petrophysical characteristics. MIP technique provide data for the determination of porosity, permeability, port type, distribution and sorting, average pore-throat radius (APR), median pore-throat radius (MPR), maximum threshold-entry radius (MTER), and other petrophysical properties (Table 1).

The objectives of this study are: (1) to develop a classification of reservoir vs. non-reservoir rocks in the Anadarko Basin based on their lithology and petrophysical characteristics; (2) to establish the relationships between various petrophysical properties; 3) to determine how these relationships are affected by sorting of the pore-throat size distribution, and (4) to build a petrophysical database for reservoir and non-reservoir rocks of the Anadarko Basin.

METHODS

We used MIP technique to analyze 91 core samples selected from wells drilled in reservoir and non-reservoir rocks of the Anadarko Basin, OK (Figure 1). The samples are identified assandstones (20), shales (24), interbedded shales and siltstones (11), siltstones

(17), grainstones (9) and lime mudstones (10), ranging in age from Devonian to Permian and in depth from 404 m to 4,975 m (Table 1).

A significant issue regarding the measurements was determining how representative the samples are for reservoir and non-reservoirs rocks of the Anadarko Basin. Based on production information included in the scout cards of each well, the sampling procedure ensured that most areas of interest were sampled. Major modification of many of the properties to be measured occurs during extraction and storage of core samples. The samples are not in situ, they reside in core warehouses, and are undoubtedly dessicated and not representative of their in situ properties. Consequently, our main concern during sampling was to select only those cores that were the freshest, thus assuring that the residing time in the core warehouse was minimal.

Following suggestions made by other researchers (e.g., Aplin et al., eds., 1999; Yang and Aplin, 2007) necessary corrections for effects of clay dessication to compensate the missing of original fluids were applied.

Before using our MIP technique, we considered alternative methods that may preserve the original fluids in place: porous plate, vapor desorption, and centrifuge methods (e.g., Newsham et al., 2004; AlHinai et al., 2008). The major drawbacks of the aforementioned methods are their longer time of experimentation (one month for a single measurement) and relatively low range of operating pressures (1000 – 1250 psi). Unlike the porous plate technique, MIP is very fast, often requiring only hours of operation rather than days or weeks (Newsham et al., 2004). In addition, MIP technique is capable of generating injection pressures up to 60,000 psi and thus allows investigating low porosity and low permeability rocks.

Petrophysical Parameters

Samples were cut from core slabs, prepared and analyzed in a Micromeritics AutoPore 9500 using the method described by Deming et al., 2002; Cranganu, 2004; Villa, 2005; and Cranganu and Villa, 2006.

The parameters apparent porosity, average pore-throat radius, median pore-throat radius and maximum threshold entry radius were determined according to Webb, 2001:

Apparent porosity, Φ (%), is measured by capillary-pressure analysis to a pressure of 60,000 psi (~414 MPa) and is defined as:

$$\Phi(\%) = \frac{V_{tot}}{V_b} \times 100 \quad (2)$$

where V_{tot} is the total intrusion volume of mercury required to fill all accessible pores of the sample and is calculated as:

$$V_{tot} = V_j \quad (3)$$

Table 1. Petrophysical data for reservoir and non-reservoir rocks from the Anadarko Basin, OK

ID	Well Name	Latitude	Longitude	Period	Formation	Depth (m)	Rock Type	MPR (µm)	APR (µm)	ϕ (%)	Kair (md)	R35 (µm)	PT	PTD	MTER (µm)	PTS	Reservoir Quality
94	VOYLES HEIRS 2	35.11	-98.86	Pennsylvanian	Springer	1497	SILTSTONE	0.170	0.018	13.20	3.057	1.12	MESO	no mode	------	PS	1
31	IDA EDWARDS 2	34.94	-98.21	Pennsylvanian	Missourian	1756	SANDSTONE	7.697	0.160	13.10	406.000	49.00	MEGA	unimodal	9.8	WS	1
63	PATTERSON UNIT D-1	35.95	-98.38	Pennsylvanian	Morrowan	2560	SHALE	2.079	0.010	12.13	131.459	11.00	MEGA	bimodal	0.003/100	PS	1
44	MALES 3-8	35.79	-99.88	Pennsylvanian	Tonkawa	2620	SANDSTONE	0.311	0.090	11.24	1.362	0.80	MESO	unimodal	0.2963	MS	1
101	WOLFE UNIT 1	35.89	-98.43	Pennsylvanian	Morrowan	2744	INTERBEDDED SH & SLST	0.024	0.013	10.00	1.051	0.76	MESO	unimodal	0.014	MS	1
95	W VERDEN HOXBAR UN 14	35.12	-98.14	Pennsylvanian	Hoxbar	2995	SANDSTONE	2.742	0.117	12.58	41.355	5.40	MACRO	unimodal	4.85	WS	1
89	TRUSPER UN 1	35.62	-98.21	Pennsylvanian	Morrowan	3192	SHALE	0.147	0.010	11.36	15.090	3.26	MACRO	bimodal	0.007/150	PS	1
52	MURPHY 1	35.33	-99.69	Pennsylvanian	Missourian	3375	SANDSTONE	0.738	0.306	14.65	6.536	1.60	MESO	no mode	------	PS	1
32	IDA EDWARDS 2	34.94	-98.21	Pennsylvanian	Missourian	1756	SHALE	0.025	0.008	6.39	7.311	3.50	MACRO	bimodal	0.004/148	MS	2
33	IDA EDWARDS 2	34.94	-98.21	Pennsylvanian	Missourian	1757	SILTSTONE	0.015	0.005	7.39	8.361	4.50	MACRO	bimodal	0.004/148	MS	2
21	HALL 2-B	35.04	-98.53	Devonian	Woodford	1890	SHALE	10.163	0.013	2.79	288.993	62.23	MEGA	unimodal	120	MS	2
55	NETTIE PATZACK 1	35.40	-97.84	Pennsylvanian	Missourian	1946	GRAINSTONE	6.761	0.015	2.13	21.390	17.00	MEGA	bimodal	0.0069/185	PS	2
35	KAMM JOHN-STON 33-1	35.47	-97.95	Pennsylvanian	Tonkawa	2085	SILTSTONE	0.300	0.008	4.88	65.235	16.00	MEGA	bimodal	0.005/149	MS	2
90	USF&G 18A-4	35.43	-97.56	Pennsylvanian	Prue	2136	SILTSTONE	0.080	0.007	4.22	32.305	12.00	MEGA	bimodal	0.004/119	MS	2
18	GARTNER 1-24	35.59	-98.75	Pennsylvanian	Prue	2174	SHALE	0.028	0.007	4.10	26.705	11.00	MEGA	bimodal	0.003/110	PS	2
41	KNIE A-1	35.23	-98.97	Pennsylvanian	Morrowan	2337	SHALE	0.012	0.008	9.88	1.937	1.10	MESO	unimodal	0.0069	MS	2
16	FARRIS UN 2-2	35.89	-99.19	Pennsylvanian	Haskell	2396	SHALE	0.034	0.008	4.51	16.749	7.70	MACRO	bimodal	0.004/140	PS	2
27	HOLM 7-1	36.14	-97.99	Devonian	Hunton	2407	LIME MUDSTONE	5.365	0.008	2.28	101.307	40.00	MEGA	bimodal	0.0034/149	PS	2
10	COULTER 1-19	36.29	-99.06	Mississippian	Chester	2411	GRAINSTONE	6.055	0.017	2.38	86.861	35.21	MEGA	unimodal	184	PS	2
15	CURTIS 2	35.40	-97.72	Devonian	Hunton	2596	SHALE	3.383	0.008	2.89	209.771	50.00	MEGA	bimodal	0.003/98	MS	2
14	CURTIS 2	35.40	-97.72	Devonian	Woodford	2603	GRAINSTONE	1.193	------	5.37	6.224	3.70	MACRO	unimodal	0.7704	WS	2

ID	Well Name	Latitude	Longitude	Period	Formation	Depth (m)	Rock Type	MPR (µm)	APR (µm)	ϕ (%)	Kair (md)	R35 (µm)	PT	PTD	MTER (µm)	PTS	Reservoir Quality
42	LOOKBAUGH 1	35.73	-98.32	Pennsylvanian	Oswego	2759	LIME MUDSTONE	0.028	0.009	5.40	10.866	5.11	MACRO	bimodal	0.005/190	PS	2
1	ARMSTRONG UNIT 2	35.33	-98.14	Pennsylvanian	Cottage Grove	2855	INTERBEDDED SH & SLST	0.262	0.015	3.07	4.111	4.70	MACRO	bimodal	0.0055/98	PS	2
83	SHAFFER 1-23	35.85	-98.23	Mississippian	Mississippian	2944	GRAINSTONE	1.811	0.024	4.26	77.955	19.98	MEGA	bimodal	0.0111/184	PS	2
13	COWAN UN 1	35.70	-98.50	Pennsylvanian	Marmaton	2967	SILTSTONE	0.014	0.007	4.82	2.210	2.21	MACRO	bimodal	0.0043/148	MS	2
74	SALLIE BLACKBEAR 1	35.78	-98.36	Pennsylvanian	Morrowan	2977	SILTSTONE	0.875	0.015	1.79	5.417	8.81	MACRO	bimodal	0.3/148	PS	2
61	NIECE 1-27	35.31	-99.30	Pennsylvanian	Des Moinesian	3075	INTERBEDDED SH & SLST	2.958	0.013	3.01	121.414	35.00	MEGA	bimodal	0.0055/148	PS	2
2	BAKER A1	35.65	-98.52	Pennsylvanian	Cherokee	3306	INTERBEDDED SH & SLST	0.489	0.012	2.97	30.122	15.60	MEGA	bimodal	0.0055/155	PS	2
87	SUMMERS 1-13	35.69	-99.60	Pennsylvanian	Marmaton	3352	SHALE	0.022	0.006	5.54	1.403	1.50	MESO	bimodal	0.003/160	PS	2
78	SCOTT-1	35.69	-99.63	Pennsylvanian	Marmaton	3355	INTERBEDDED SH & SLST	1.950	0.009	2.90	24.196	14.00	MEGA	bimodal	0.0034/154	PS	2
80	SCOTT-1	35.69	-99.63	Pennsylvanian	Marmaton	3355	SHALE	0.022	0.007	5.50	8.029	4.21	MACRO	bimodal	0.003/180	PS	2
49	MERRICK 1	35.76	-99.61	Pennsylvanian	Skinner	3419	INTERBEDDED SH & SLST	0.039	0.010	5.27	7.119	4.07	MACRO	bimodal	0.0087/121	PS	2
25	HOLLAND 2	34.76	-97.75	Pennsylvanian	Morrowan	3496	GRAINSTONE	0.372	0.030	4.11	5.212	4.20	MACRO	no mode	------	PS	2
46	MAYFIELD 1-34	35.39	-99.83	Pennsylvanian	Missourian	3529	GRAINSTONE	0.272	0.030	4.47	5.198	3.90	MACRO	no mode	------	PS	2
60	NIECE 1-27	35.31	-99.30	Pennsylvanian	Des Moinesian	3594	GRAINSTONE	28.430	0.016	1.11	91.122	70.00	MEGA	unimodal	97	MS	2
68	RALPH CRALL UNIT 1	35.56	-98.69	Pennsylvanian	Atoka	3972	SHALE	0.250	0.016	7.76	154.013	17.76	MEGA	bimodal	0.001/150	PS	2
70	RUTH 1-28	35.57	-99.54	Pennsylvanian	Red Fork	3981	INTERBEDDED SH & SLST	2.648	0.014	6.27	183.646	23.68	MEGA	bimodal	0.0055/149	PS	2
85	SPIEKER	35.44	-99.27	Pennsylvanian	Red Fork	4133	INTERBEDDED SH & SLST	0.032	0.011	4.46	5.363	3.98	MACRO	bimodal	0.001/97	PS	2
5	BEUTLER 1-13	35.52	-99.37	Pennsylvanian	Red Fork	4138	SILTSTONE	0.050	0.050	7.00	6.528	4.00	MACRO	bimodal	0.003/149	MS	2
6	BEUTLER 1-13	35.52	-99.37	Pennsylvanian	Red Fork	4139	SHALE	0.040	0.007	6.00	5.128	3.00	MACRO	bimodal	0.003/180	PS	2
47	MCCALLA RANCH 1-12	35.10	-97.79	Devonian	Woodford	4150	SANDSTONE	1.627	------	6.52	61.219	12.00	MEGA	bimodal	0.05/70	PS	2

Table 1. (Continued)

ID	Well Name	Latitude	Longitude	Period	Formation	Depth (m)	Rock Type	MPR (μm)	APR (μm)	ϕ (%)	Kair (md)	R35 (μm)	PT	PTD	MTER (μm)	PTS	Reservoir Quality
48	MCCALLA RANCH 1-12	35.10	-97.79	Devonian	Woodford	4151	SILTSTONE	1.257	------	5.52	60.919	10.00	MEGA	bimodal	0.37545/63	WS	2
36	KENNEMER 1-22	35.24	-99.41	Pennsylvanian	Des Moinesian	4201	LIME MUDSTONE	14.356	0.024	2.81	168.272	45.00	MEGA	bimodal	0.004/160	PS	2
102	YOUNG-1	35.81	-99.74	Pennsylvanian	Morrowan	4438	SANDSTONE	0.326	------	3.78	2.690	3.06	MACRO	unimodal	0.1153	WS	2
24	HARRIS 1	34.85	-97.84	Mississippian	Sycamore	4515	LIME MUDSTONE	0.045	0.024	3.30	20.444	11.34	MEGA	bimodal	0.0224/120	PS	2
37	KENNEMER 1-22	35.24	-99.41	Pennsylvanian	Des Moinesian	4575	SHALE	2.996	0.010	4.67	220.366	34.00	MEGA	bimodal	0.004/110	PS	2
66	PECK 1-21	35.50	-99.32	Unknown	Unknown	4592	LIME MUDSTONE	0.021	0.008	4.13	26.993	11.00	MEGA	bimodal	0.006/160	PS	2
76	SCHMDT 1	35.15	-99.02	Pennsylvanian	Morrowan	4975	SHALE	1.646	0.014	7.75	176.294	19.25	MEGA	bimodal	0.008/130	PS	2
88	TAYLOR-1	35.52	-99.60	Pennsylvanian	Pennsylvanian	unknown	LIME MUDSTONE	0.014	0.005	4.43	5.265	3.96	MACRO	bimodal	0.002/165	PS	2
96	WEATHERLY 1	35.55	-99.41	Pennsylvanian	Morrowan	unknown	SHALE	0.242	0.013	6.55	102.678	16.20	MEGA	bimodal	0.009/110	PS	2
28	HUBERT RAPP 1-32	36.17	-99.04	Pennsylvanian	Morrowan	1463	SHALE	0.014	0.009	9.10	0.004	0.03	NANO	unimodal	0.0087	MS	3
56	NETTIE PATZACK 1	35.40	-97.84	Pennsylvanian	Missourian	1952	SILTSTONE	0.008	0.006	4.17	0.021	0.16	MICRO	unimodal	0.0043	MS	3
7	BORDEN-1	36.20	-99.32	Pennsylvanian	Pennsylvanian	2075	SANDSTONE	0.014	0.010	5.82	0.018	0.11	MICRO	unimodal	0.0069	MS	3
11	COULTER 1-19	36.29	-99.06	Mississippian	Chester	2411	SILTSTONE	0.055	0.017	7.14	0.084	0.23	MICRO	unimodal	0.057	MS	3
92	VIERSEN UN B1	35.96	-98.40	Pennsylvanian	Morrowan	2490	INTERBEDDED SH & SLST	0.025	0.015	8.45	0.008	0.05	NANO	unimodal	0.0224	MS	3
45	MALES 3-8	35.79	-99.88	Pennsylvanian	Tonkawa	2620	SILTSTONE	0.027	0.010	4.34	0.063	0.30	MICRO	unimodal	0.0069	MS	3
100	WINTERS GAS UN 1	35.98	-98.48	Pennsylvanian	Morrowan	2663	INTERBEDDED SH & SLST	0.014	0.008	7.96	0.297	0.44	MICRO	unimodal	0.0055	MS	3
65	PAYNE 17-1	35.34	-97.73	Mississippian	Mississippian	2713	SHALE	0.009	0.007	6.77	0.002	0.03	NANO	unimodal	0.0055	WS	3
75	SCHAFER 1	36.02	-98.68	Pennsylvanian	Morrowan	2793	SANDSTONE	0.060	0.052	9.10	0.016	0.07	NANO	unimodal	0.057	WS	3

ID	Well Name	Latitude	Longitude	Period	Formation	Depth (m)	Rock Type	MPR (µm)	APR (µm)	ϕ (%)	Kair (md)	R35 (µm)	PT	PTD	MTER (µm)	PTS	Reservoir Quality
17	FRICK 1	34.81	-97.80	Pennsylvanian	Cunningham	2931	SANDSTONE	0.025	0.013	6.83	0.038	0.15	MICRO	unimodal	0.0111	MS	3
72	SALLIE BLACKBEAR 1	35.78	-98.36	Pennsylvanian	Atoka	2961	LIME MUDSTONE	0.008	0.008	6.37	0.0003	0.01	NANO	unimodal	0.0069	WS	3
69	REAGAN #1	35.95	-99.37	Pennsylvanian	Oswego	2987	LIME MUDSTONE	0.007	0.006	4.63	0.032	0.19	MICRO	unimodal	0.0034	MS	3
73	SALLIE BLACKBEAR 1	35.78	-98.36	Pennsylvanian	Morrowan	2994	SANDSTONE	0.256	0.044	8.72	0.511	0.56	MESO	unimodal	0.2336	MS	3
30	HUNT UNT 1	35.57	-98.14	Pennsylvanian	Atoka	3120	SHALE	0.007	0.007	8.20	0.029	0.11	MICRO	unimodal	0.0055	WS	3
20	HALL 1-11	36.14	-99.41	Pennsylvanian	Morrowan	3167	SHALE	0.006	0.006	5.90	0.0004	0.01	NANO	unimodal	0.0055	WS	3
51	MINTON 1	35.89	-98.76	Mississippian	Chester	3176	GRAINSTONE	0.265	0.155	9.54	0.069	0.38	MICRO	unimodal	0.2951	MS	3
50	MINTON 1	35.89	-98.76	Mississippian	Chester	3185	LIME MUDSTONE	0.020	0.012	6.13	0.158	0.16	MICRO	unimodal	0.0111	MS	3
43	LYON UNIT B-1	35.67	-98.32	Pennsylvanian	Morrowan	3231	SILTSTONE	0.020	0.010	8.19	0.005	0.04	NANO	unimodal	0.0111	MS	3
38	KIPPENBERGER 1-23	35.76	-98.66	Pennsylvanian	Morrowan	3261	GRAINSTONE	0.203	0.045	9.56	0.359	0.42	MICRO	unimodal	0.2336	MS	3
57	NEWTON-JOHN 1	34.83	-97.84	Pennsylvanian	Primrose	3283	SANDSTONE	0.033	0.022	7.74	0.242	0.40	MICRO	unimodal	0.014	MS	3
77	SCOTT-1	35.02	-98.00	Pennsylvanian	Marchand	3316	SANDSTONE	0.112	0.069	5.41	0.048	0.21	MICRO	unimodal	0.0912	WS	3
58	NEWTON-JOHN 1	34.83	-97.84	Pennsylvanian	Primrose	3317	SILTSTONE	0.030	0.014	8.06	0.415	0.53	MESO	unimodal	0.0224	WS	3
67	PHIPPS 1	34.73	-97.73	Pennsylvanian	Morrowan	3330	SANDSTONE	0.034	0.024	6.48	0.012	0.08	NANO	unimodal	0.0177	MS	3
86	SUMMERS 1-13	35.69	-99.60	Pennsylvanian	Marmaton	3352	SANDSTONE	0.168	0.043	9.20	0.096	0.20	MICRO	unimodal	0.1846	WS	3
93	VINCENT 1	36.18	-99.96	Pennsylvanian	Morrowan	3383	SHALE	0.020	0.011	7.23	0.437	0.60	MESO	unimodal	147	MS	3
79	SCOTT-1	35.69	-99.63	Pennsylvanian	Marmaton	3383	SHALE	0.007	0.006	4.75	0.033	0.19	MICRO	unimodal	0.0034	WS	3
26	HOLLAND 2	34.76	-97.75	Pennsylvanian	Morrowan	3496	SANDSTONE	0.173	0.037	5.32	0.780	1.10	MESO	unimodal	0.1152	WS	3
22	HARGRAVE-1	35.56	-98.62	Pennsylvanian	Red Fork	3508	SILTSTONE	0.009	0.008	3.06	0.001	0.03	NANO	unimodal	0.0055	WS	3
3	BERRYMAN 1-R	35.96	-99.71	Pennsylvanian	Morrowan	3824	SANDSTONE	0.065	0.027	7.37	0.050	0.17	MICRO	unimodal	0.0224	MS	3
4	BEUTLER 1-13	35.52	-99.37	Pennsylvanian	Red Fork	4138	SANDSTONE	0.032	0.026	6.16	0.040	0.17	MICRO	unimodal	0.0224	WS	3
34	JONES HUNTON-1	35.73	-98.83	Devonian	Woodford	4282	LIME MUDSTONE	0.488	0.283	6.53	0.454	0.67	MESO	unimodal	0.4754	WS	3
9	CORE HOLE 2-Z	35.05	-99.04	Permian	Permian	404	SANDSTONE	0.016	0.011	13.47	0.003	0.02	NANO	unimodal	0.0177	WS	4

Table 1. (Continued)

ID	Well Name	Latitude	Longitude	Period	Formation	Depth (m)	Rock Type	MPR (µm)	APR (µm)	ϕ (%)	Kair (md)	R35 (µm)	PT	PTD	MTER (µm)	PTS	Reservoir Quality
12	COVINGTON 1	35.05	-99.67	unknown	Dolomite	529	SANDSTONE	0.032	0.021	11.01	0.008	0.04	NANO	unimodal	0.0357	WS	4
82	SELBY 51	34.44	-97.69	Permian	Permian	1454	SHALE	0.008	0.008	10.44	0.001	0.01	NANO	unimodal	0.0087	WS	4
54	MYIA 1-11	34.66	-97.70	Pennsylvanian	Morrowan	2019	INTERBEDDED SH & SLST	0.011	0.008	11.29	0.002	0.02	NANO	unimodal	0.0087	WS	4
59	NICHOLS ROYALITY 1	34.89	-98.03	Pennsylvanian	Pennsylvanian	2314	SANDSTONE	0.125	0.029	13.63	0.171	0.20	MICRO	unimodal	0.2342	WS	4
40	KNIE A-1	35.23	-98.97	Pennsylvanian	Morrowan	2338	SILTSTONE	0.046	0.017	10.07	0.219	0.30	MICRO	unimodal	0.0111	MS	4
91	VIERSEN UN B1	35.96	-98.40	Pennsylvanian	Atoka	2476	SILTSTONE	0.016	0.012	11.33	0.003	0.02	NANO	unimodal	0.0177	WS	4
97	WILEY UN 1	35.99	-98.48	Pennsylvanian	Atoka	2626	SHALE	0.011	0.009	10.14	0.196	0.28	MICRO	unimodal	0.0088	WS	4
98	WILEY UN 1	35.99	-98.48	Pennsylvanian	Atoka	2627	SILTSTONE	0.021	0.011	10.44	0.1206	0.26	MICRO	unimodal	0.0088	WS	4
39	KIPPENBERGER # 1-23	35.76	-98.66	Pennsylvanian	Morrowan	3261	SHALE	0.012	0.009	10.55	0.656	0.55	MESO	unimodal	0.0069	WS	4

ID = the numbers are the same as those in Figure 1; MPR = Median Pore-throat Radius; APR = Average Pore-throat Radius; ϕ =porosity; Kair = air permeability; R35 = Pore-throat Radius at 35% mercury intrusion; PT = Port Type; PTD = Pore-throat Size Distribution; MTER = Maximum Threshold Entry Radius; PTS = Pore-throat Size Sorting.

Table 2. Statistical correlations for various petrophysical parameters for reservoir and non-reservoir rocks

	Reservoir rocks				Non Reservoir Rocks			
	Φ vs. Kair	Φ vs. APR	Φ vs. MPR	Φ vs. MTER	Φ vs. Kair	Φ vs. APR	Φ vs. MPR	Φ vs. MTER
r	0.050	0.587	-0.267	-0.689	0.053	-0.116	-0.042	-0.047
p	0.731	0.000*	0.061	0.025*	0.740	0.469	0.793	0.771

* Statistical significant correlations exists between parameters; $p < 0.05$

Here, the jth data point (measurement step) is the first such that:

$$P_{j+1} \leq P_j - 10 \text{ and } P_{j+1} \leq P_j \times 0.995 \tag{4}$$

where P_j is the head-corrected pressure as stored by the instrument.

V_b is the bulk volume of mercury at atmospheric pressure and is calculated as:

$$V_b = V_p - V_m \tag{5}$$

where V_p is user-entered volume of penetrometer and V_m is the volume of mercury in penetrometer.

Average Pore-Throat Radius, APR (μm). The idealized cylindrical geometry of the pore throats in the system is a function of the radius of the cross sectional circular area of the pore throat. This average value, expressed in microns, is calculates as follows:

$$\text{APR}(\mu\text{m}) = 2 \times \frac{I_{total}}{A_{total}} \tag{6}$$

where I_{tot} is the total specific intrusion volume of sample and A_{tot} is the total specific pore area of sample. Both parameters are calculated by the instrument.

Median Pore-Throat Radius, MPR (μm), is calculated as the arithmetic average of the median pore radius by volume (MPRV) and the median pore radius by area (MPRA).

MPRV is calculated as follows:

$$\text{MPRV} = \text{MPRV}_\text{k} \tag{7}$$

where MPRV_k is calculated from P_k (head-corrected pressure as stored), and P_k is interpolated from I_k and the collected data. I_k is defined as:

$$I_k = \frac{I_{tot}}{2}$$
$$I_{tot} = \frac{V_{tot}}{W_s} \tag{8}$$

where V_{tot} is defined by Equation (3) and W_s is user-entered sample weight.

MPRA is calculated as follows:

$$MPRA = MPRA_k \tag{9}$$

where $MPRA_k$ is calculated from Pk (head-corrected pressure as stored), and Pk is interpolated from A_k and the collected data. A_k is defined as:

$$A_k = \frac{A_{tot}}{2} \tag{10}$$

where A_{tot} is the total specific pore area.

Commonly, MPRV and MPRA differ, because the smaller pores contribute more to the total pore surface area than do larger pores for a given increment of mercury imbibition. Therefore, the area distributions have a tendency to shift to the smaller pore sizes where compared to pore volume distributions (Tanguay and Friedman, 2001).

Maximum Threshold-Entry Radius, MTER (μm) is the entry radius at which significant invasion of a sample with mercury occurs (Tanguay and Friedman, 2001). The recognition of this parameter is based on a histogram (Figure 2) created with the incremental percentage of intrusion (times 10 to emphasize the values) and the size of the pore throat radius to which it is equivalent. MTER is the largest radius of the largest population of ubiquitous pore-throat sizes in a sample. The peak or peaks on the histogram reflect the distribution of the different pore-throat radius populations in a rock sample (Pore-Throat Size Distribution or PTD). One peak represents a unimodal distribution; two peaks represent a bimodal distribution, and multiple peaks represent a multiple (no mode) distribution (Figure 3).

These distributions are linked to the locations of voids in the rock and they are important in determining if one, two, or more pore-throat systems dominate the pore geometry of a geologic formation. In turn, this determination has consequences in recognizing the presence of one significant fluid invasion (unimodal distribution), two significant fluid invasions (bimodal distribution0, or no significant fluid invasion (no mode distribution).

As suggested by Hartmann (pers. comm., 2005), the distributions of pore throats represented by more than one mode are due to either the sample having two different pore throat types corresponding to laminations or to the presence of connected vugs in a microcrystalline dolomite or calcite matrix, or grainstone, where the intergranular is the largest port size, and micritized intergranular (microcrystalline) is the smallest pore throat size.

Pore-Throat Sorting, PTS, is a measure of the sorting of the sizes of the pore throats in a sample. Measurements of PTS from intrusion capillary curves have been both qualitative and quantitative. Jennings (1987) proposed the following formula for PTS:

$$PTS = \left[\frac{3^{rd}\ \text{Quartile Pressure}}{1^{st}\ \text{Quartile Pressure}} \right]^{1/2} \tag{11}$$

where the first and third-quartile pressures are obtained directly from the capillary pressure curve and reflect the 25 and 75% mercury saturation pressures adjusted for irreducible saturation. A PTS value of 1.0 represents a perfectly horizontal plateau, while values much above 5.0 pertain to curves displaying little or no plateau development.

Hartmann and Beaumont (1999) and Tanguay and Friedmann (2001) measured the sorting of pore throats by visual graphic analysis of the steepness of the plateau of the capillary curves. In general, a horizontal plateau indicates a well sorting of the pore throats (Figure 4). The values used are PS - Poorly sorting; MS - Medium Sorting, and WS - Well Sorting. According to Tanguay and Friedman (2001), well-sorted pore-throat sizes (WS) are characterized graphically by a MTER at less than 20% mercury saturation, along with a horizontal to sub-horizontal plateaued injection curve that has a unimodal pore-throat size distribution. Capillary-pressure curves of pore throats that have a moderately sorting size distribution (MS) are generally sinusoidal in shape and have a MTER between 10 and 40% mercury saturation. Finally, capillary-pressure curves of pore throats that have a poorly sorted pore-throat size distribution (PS) are generally oblique or diagonal and have little to no plateau and a poorly defined MTER.

The significance of PTS centers on the rock's ability to accept oil saturation. Jennings (1987, p. 1199) described the petrophysical role of pore throat sorting in reservoir analysis in these terms: "In well-sorted rocks, once a threshold buoyancy pressure is obtained, oil will rapidly saturate the porosity up to the maximum capacity. Poorly sorted rocks require a pressure increase over a much broader range to obtain the same level of oil saturation."

Pore-Throat Radius at 35% mercury intrusion, R35 (μm), is a concept developed by Winland (1972, 1976) to empirically relate apparent porosity Φ, air permeability Kair (md), and pore throat radius at 35% of intrusion volume R35 (μm) (Figure 2)[10]. The relationship is

$$\log R35 = 0.732 + 0.588 \log Kair - 0.864 \log \Phi \tag{12}$$

A graphical solution of this equation is illustrated in Figure 5. Using R35 instead of the Kair/Φ ratio for characterizing rock quality of water-wet rocks has several advantages (Hartmann and Beaumont, 1999):

- R35 is an understandable number;
- R35 can be determined from capillary pressure analysis and related to Kair/Φ values;

[10] It should be emphasized that Winland equation (12) was developed by using mostly sandstone and carbonate rock samples. This means that air permeability Kair for other rocks (e.g., shales) should be considered as informative.

- If two variables are known (Kair, Φ, or R35), then the other variable can be calculated using Equation 12 or estimated from a Kair/Φ plot with R35 contours.

The R35 value of a given rock type both reflects its depositional and diagenetic fabric and influences fluid flow and reservoir performance (Hartmann and Coalson, 1990). Consequently, estimating R35 directly from capillary pressure data provides the background for a common zonation that can be used by both geologists and reservoir engineers. It also offers a basis for defining reservoir flow units (see next).

Port Type (PT) concept was used by Coalson et al. (1985) as a link between the R35 values and characterization of the pore systems by size. Pore systems in reservoir rocks are characterized by port types, which are measurable characteristics that can represent the reservoir quality. Extending Martin et al. (1997) classification, five petrophysical flow units with different reservoir performances are distinguished by ranges of R35 (Figures 2 and 3):

(1) *Megaport* – flow units are defined as having an R35 ranging above a threshold of 10 μm. Production of medium-gravity crudes can readily attain tens of thousands of barrels per day from a megaport flow unit if zonal thickness and other factors are constant.
(2) *Macroport* – flow units having slightly smaller pore throat sizes that, with all other constraints held constant, are capable of thousands of barrels of oil per day. These flow units are defined as having an R35 ranging between 2 and 10 μm.
(3) *Mesoport* – flow units having an R35 ranging between 0.5 – 2 μm. These units may allow only hundreds of oil day with all other factors held constant.
(4) *Microport* – flow units having an R35 ranging between 0.1 – 0.5 μm. Although numerous tight gas reservoirs have these R35 properties, microport flow units are mostly non-reservoir zones. Wells with mostly microport flow units produce at best a few barrels of oil per day on pump.
(5) *Nanoport* – flow units that have an R35 ranging of less than 0.1 μm. The nanoport flow units best characterize non-reservoir zones and are of far more interest as potential seals for higher quality reservoir downdip.

Reservoir Quality. Using cut off values of 1 md for permeability and 10% for apparent porosity (values commonly used in the Anadarko Basin), we separated the plot in Figure 5 into four areas called windows as follows:

Window 1 - defined by apparent porosity ranging from 10% to 20% and air permeability ranging between 1 md - 10 d; port types are mega, macro, and meso;
Window 2 – defined as having apparent porosity between 0% - 10% and air permeability between 1 md - 10 d; port types are mega, macro, and meso;
Window 3 – defined by apparent porosity ranging from 0% to 20% and air permeability ranging between 0.0001 md – 1 md; port types are micro and nano;
Window 4 - defined as having apparent porosity between 10% - 20% and air permeability between 0.0001 md – 1 md; port types are micro and nano.

Windows 1 and 2 would represent reservoir rocks, while windows 3 and 4 would represent non-reservoir rocks.

Sample Lithology

The samples used in this research were classified as follows:

Shales: The 24 samples display very fine grain sizes. Micaceous material is present in some samples. The color varies from light gray to black. Samples exhibit fissibility and laminations.

Sandstones: The 20 samples display medium to fine grain sizes. The color varies from white to light gray. Samples are well indurated and massive. Quartz and feldspars are abundant.

Lime mudstones: The 10 samples display clay-size carbonate particles. They are finely crystalline and react with HCl. The color varies from light gray to dark gray. In general, the samples are very well indurated. Samples contain less than 10% grains and are mud supported.

Grainstones: The 9 samples are grain supported carbonate rocks with no mud. They react with HCl. The color varies from light gray to gray. Samples are in general massive and very well indurated.

Siltstones: The 17 samples display grains sizes smaller than 62 microns, lack the fissibility and laminations of shales. Similar in color to shales.

Interbedded shales and siltstones: The 11 samples exhibit planar bedding. The samples are very thinly laminated to thickly laminated. The different beds display grain sizes varying form very fine sand, silty sand, silty clay to clay. Their color varies from gray to light gray. They are well indurated. If a fluid flow occurs along the interbedded shales and siltstones, then the permeability would depend on the flow direction (anisotropic effect).

RESULTS AND DISCUSSION

MIP technique was previously used in the Anadarko Basin to investigate the physical properties of clastic reservoir rocks (Keighin, 1997), the seal capacity of nonsmectite shales (Krushin, 1997) or to estimate the capillary sealing (Deming et al., 2002; Cranganu, 2004; Cranganu and Villa, 2006). Here we report nine petrophysical parameters: apparent porosity, $\Phi(\%)$; median pore-throat radius, MPR (μm); average pore-throat radius, APR (μm); calculated air permeability, Kair (md); port types (PT); pore-throat sorting, PTS; pore-throat distribution; maximum threshold entry radius, MTER (μm); and reservoir quality (Figures 2 - 5). Other measured parameters (bulk density, skeleton density, entry pressure, threshold pressure, displacement pressure, curve types and shapes, and sealing capacity of non-reservoir rocks have been determined, but not reported here).

Ninety-one samples have been lithologically identified and measured (Table 1). Figure 1 shows their locations. The coverage of the basin by the selected samples is quite uniform. The depth of the samples ranges between 404 and 4,975 m. Stratigraphically, the samples are of Devonian (6), Mississippian (7), Pennsylvanian (76), and Permian (2) ages.

Twenty samples were identified as sandstones, 9 as grainstones, 11 as interbedded shales and siltstones, 24 as shales, 10 as lime mudstones, and 17 as siltstones.

Reservoir vs. Non-Reservoir Rocks

Based on R35 values and cut off values of 1 md for permeability and 10% for porosity (values widely used in the Anadarko Basin petroleum business), we classified the samples as reservoir vs. non-reservoirs rocks by using the graphical solution of Equation 12 (Figure 5). There are 50 reservoir samples (8 in window 1 and 42 in window 2) and 41 non-reservoir samples (31 in window 3 and 10 in window 4). The inclusion of samples of window 2 among the reservoir rock, although they have porosity values smaller than 10%, is motivated by the fact that those samples exhibit port types larger than 0.5 μm (meso, macro and mega sizes). Similarly, samples from window 4 have been classified as non-reservoir because they exhibit port types smaller than 0.5 μm (mostly micro and nano sizes), despite the fact that they have porosities larger than 10%.

Based on the data in Table 1 and following the classifications illustrated in Figures 2 - 5, we determined the following distribution of petrophysical parameters for the investigated reservoir and non-reservoir rocks of the Anadarko Basin (Table 3):

APR, MTER, and porosity values are relatively comparable for reservoir and non-reservoir rocks, thus making them unsuitable for discrimination between the two rock types (Figures 9 and 10). This situation most likely occurs because lithological heterogeneity renders non-distinct values of these three parameters.

Table 3. Distribution of petrophysical parameters for the investigated reservoir and non-reservoir rocks of the Anadarko Basin, OK

	Reservoir Rocks (Windows 1 and 2, Figures 5, 6A, and 6B)	Non-reservoir Rocks (Windows 3 and 4, Figures 5, 6C, and 6D)
MPR (μm)	0.012 - 28.43	0.008 - 0.488
APR (μm)	0.005 - 0.306	0.006 – 0.283
Φ (%)	1.11 – 14.65	3.06 – 13.63
Kair (md)	1.051 – 406	0.0003 – 0.780
R35 (μm)	0.76 – 70	0.001 – 1.1
PTD	bimodal (36), unimodal (10), no mode (4 samples)	unimodal (all 41 samples)
MTER(μm)	0.0014 - 184	0.0034 - 147
PTS	PS-33; MS-12; WS-5	MS-19; WS-22.
PT	mega-25; macro-19; meso-6	meso-6; micro-21; nano-14

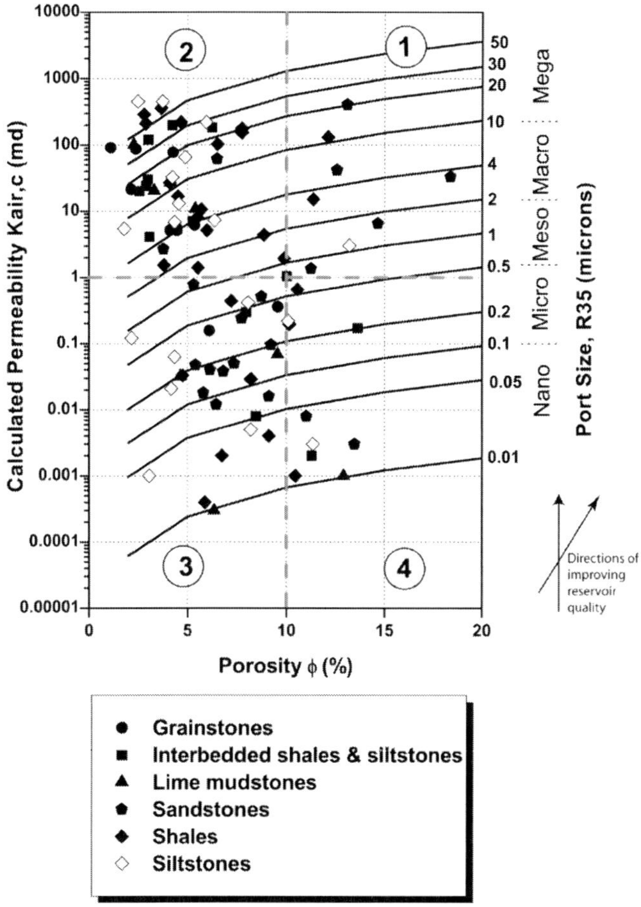

Figure 5. Reservoir quality areas (windows) for analyzed samples from the Anadarko Basin, OK. Reservoir rocks: windows 1 and 2. Non-reservoir rocks: windows 3 and 4. The arrows indicate directions of improving the reservoir quality. The best reservoir rocks are found in window 1 and the worst in window 3.

MPR values are almost 70 times larger for reservoir (0.012 – 28.43 μm) than for non-reservoir rocks (0.008 – 0.488 μm), while Kair and R35 values are each at least one order of magnitude larger in the case of reservoir rocks (Kair: 1.051 – 406 md vs. 0.0003 – 0.780 md; R35: 0.76 – 70 μm vs. 0.001 – 1.1 μm) (Figures 9 and 10). We hypothesize that the presence of reservoir samples with mostly mega and macro port types may explain the net distinction from the non-reservoir rocks, which have mostly micro and nano port types.

Another petrophysical distinction between the two types of rocks is provided by the pore-throat size distribution (PTD). It usually indicates the number of various significant fluid intrusions (one, two, or none). The reservoir rocks display a majority (36 out of 50 samples) of bimodal pore-throat distribution, implying that two significant fluid invasions may be present at different pore-throat sizes (Figure 3). In contrast, all non-reservoir rocks samples displays only unimodal distribution of their pore throat systems (Figure 3). This situation may be related to PTS: reservoir rocks are dominated by poorly (33 samples) and medium sorting (12), while the non-reservoir rocks contain predominantly pore throats with well sorting (22 samples) and medium sorting (19).

Relationships between Various Petrophysical Parameters and Lithology

Petrophysical characteristics are influenced by lithology, as evident in Table 1. Lateral and vertical change in the formation's mineral composition, with inherent grain fabric, is the primary factor that controls the lateral and vertical variations in petrophysical parameters of the reservoir and non-reservoir rocks, as well as the distinct shapes of the capillary-pressure curves (Figures 2 – 5).

A

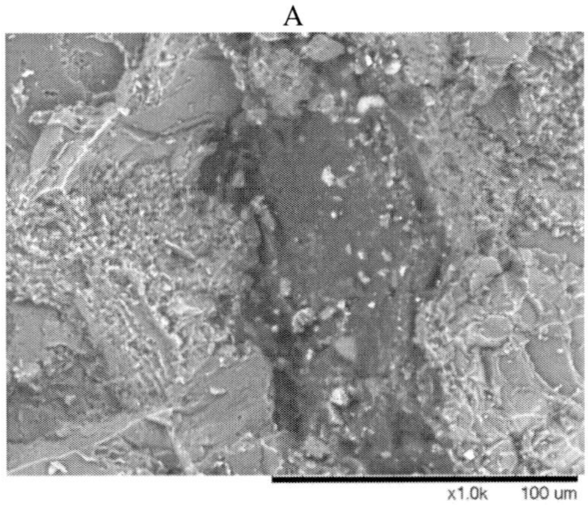

B

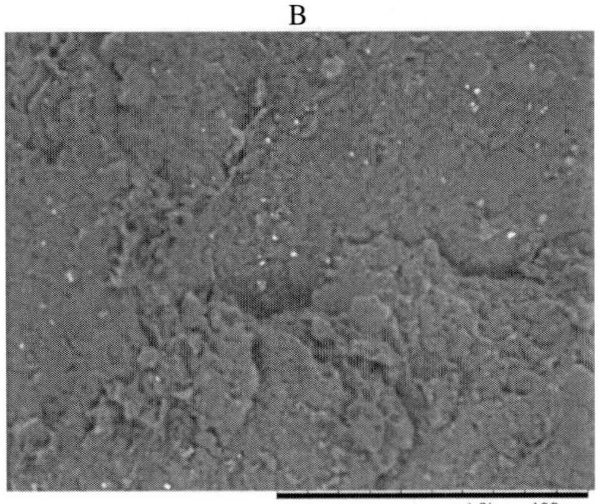

C

Figure 6. (continued).

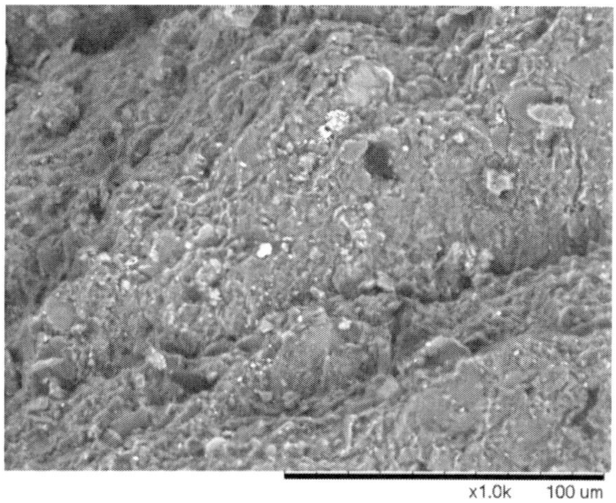

D

Figure 6. SEM photographs of representative samples from the four quality reservoir windows separated in Figure 5.
A (window 1): sample ID# 32, Ida Edwards 2 well, depth 1,756 m, Pennsylvanian (Missourian) sandstone; porosity: 13.10%; air permeability: 406 md; mega port type; unimodal pore-throat distribution; well sorted pore throats;
B (window 2): sample ID# 60, Niece 1-27 well, depth 3,594 m, Pennsylvanian (Des Moinessian) grainstone; porosity: 1.11%, air permeability 91.122 md, mega port type, unimodal pore-throat distribution, moderately sorted pore throats;
C (window 3): sample ID# 65, Payne 17-1 well, depth 2,713 m, Mississippian shale; porosity: 6.77%; air permeability; 0.002 md; nano port type; unimodal pore throat distribution; well sorted pore throats;
D (window 4): sample ID# 41, Kippenberger 1-23 well, depth 3,261 m, Pennsylvanian (Morrowan) shale; porosity: 10.55%; air permeability; 0.656 md; meso port type; unimodal pore-throat distribution; well sorted pore throats.

Figure 7 further illustrates this relationship: the majority of samples that have WS PTS are sandstones (44%); most samples that exhibit MS PTS are siltstones (32%), whereas most samples with PS PTS are shales (33%). The distributions are significantly different (χ^2 = 23.606, DF = 10, p-value = 0.009). Given that PTS is linked to the reservoir characterization, it can be consequently assumed that lithology is also related to reservoir characterization.

Porosity Relationships

Table 2 contains the Pearson correlation coefficients (r) and p-values for correlations between porosity (Φ and air permeability (Kair), average pore-throat radius (APR), median pore-throat radius (MPR), and maximum threshold entry radius (MTER). The values are indicated separately for reservoir and non-reservoir rocks.

Only two correlations (Φ vs. APR and Φ vs. MTER, both for reservoir rocks) have a statistical significance with p-value <0.05. We assume that lithological heterogeneity, depositional environments, lateral and vertical change in the formation's mineral composition, with inherent grain fabric, may be responsible for such situations. It is also possible that these correlations are attributable to the existing port types (mostly mega and macro sizes for reservoir rocks, while mostly micro and nano sizes for non-reservoir rocks). Similarly, the lack of any correlation between porosity and other petrophysical parameters in non-reservoir rocks may be caused by the predominance of micro and nano port types.

Composite plots of apparent porosity (Φ) (%) vs. calculated air permeability Kair (md), average pore-throat radius APR (μm), median pore-throat radius MPR (μm), and maximum threshold entry radius MTER (μm) for various lithological types (Figure 8) and reservoir quality (Figure 9) do not reveal particular patterns. We consider that heterogeneous composition, various depositional environments and tectonic settings have influenced the distributions depicted in those two figures. For example, apparent porosity varies between 1.11 - 9.56% for grainstones; 2.90 - 11.29% for interbedded shales and siltstones; 2.28 - 9.54% for lime mudstones; 3.78 - 13.47 for sandstones; 2.79 - 12.13 for shales; and 5.41 - 13.20% for siltstones. Average pore-throat radii are, in general, one order of magnitude smaller than median pore-throat radii (0 -0.3 μm vs. 0 - 30 μm). MTER values are mostly grouped in the interval 0.001 - 1 μm, with few values larger than this.

We further statistically analyzed (Figure 10) the main petrophysical parameter relationships with the reservoir quality.

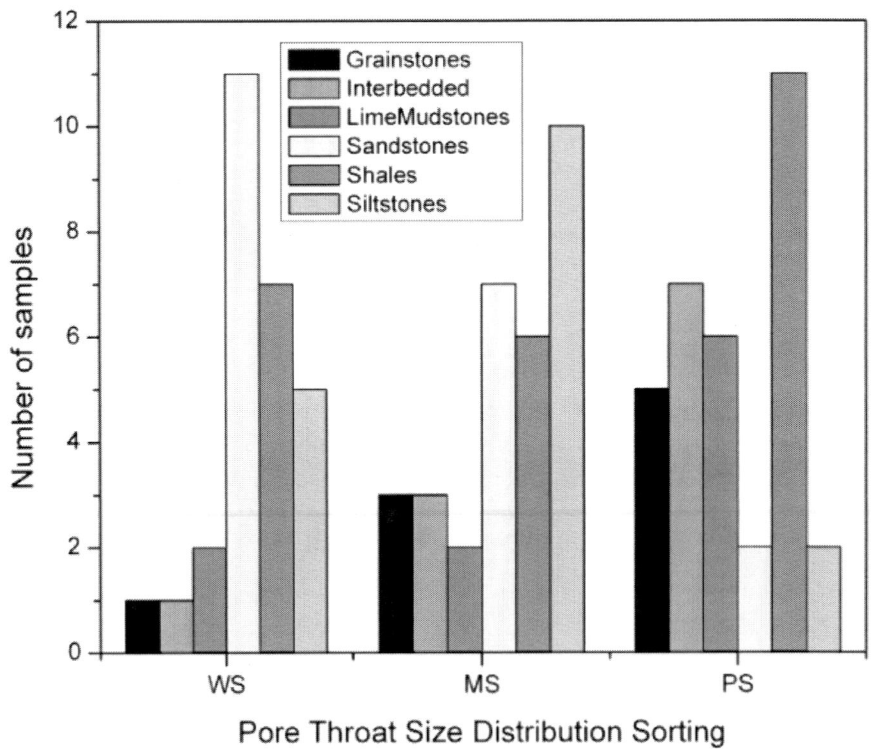

Figure 7. The distribution of grainstone, interbedded shale and siltstone, lime mudstone, sand-stone, shale, and siltstone analyzed samples that contain well-sorted (WS), medium sorted (MS), or poorly sorted (PS) pore throats (PTS). The link between lithology and pore-throat sorting is illustrated by the fact that the most samples with WS PTS are sandstones (44%); most samples that exhibit MS PTS are siltstones (32%), whereas most samples with PS PTS are shales (33%). The distributions are significantly different (χ^2 = 23.606, DF = 10, p-value = 0.009).

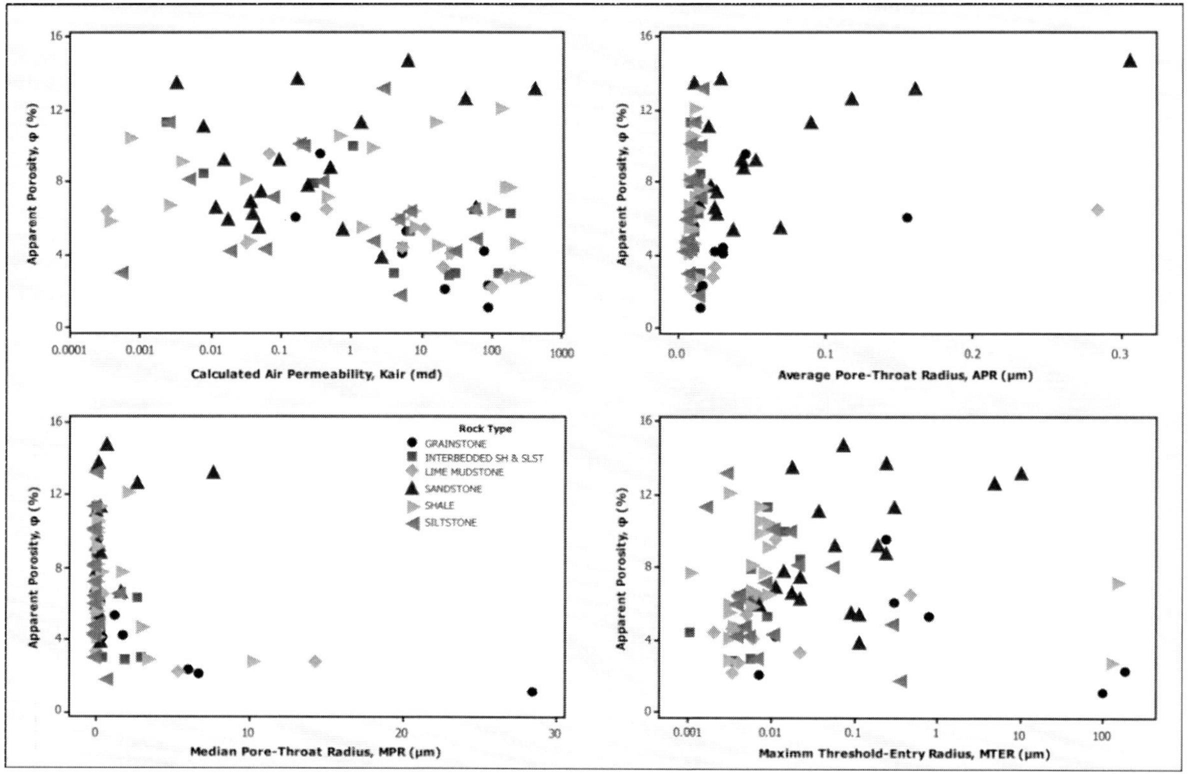

Figure 8. Composite plot of apparent porosity, ϕ (%) vs. air permeability, Kair (md), average pore-throat radius, APR (μm), median pore-throat radius, MPR (μm), and maximum threshold entry radius, MTER (μm) for various lithological types. Details are given in text.

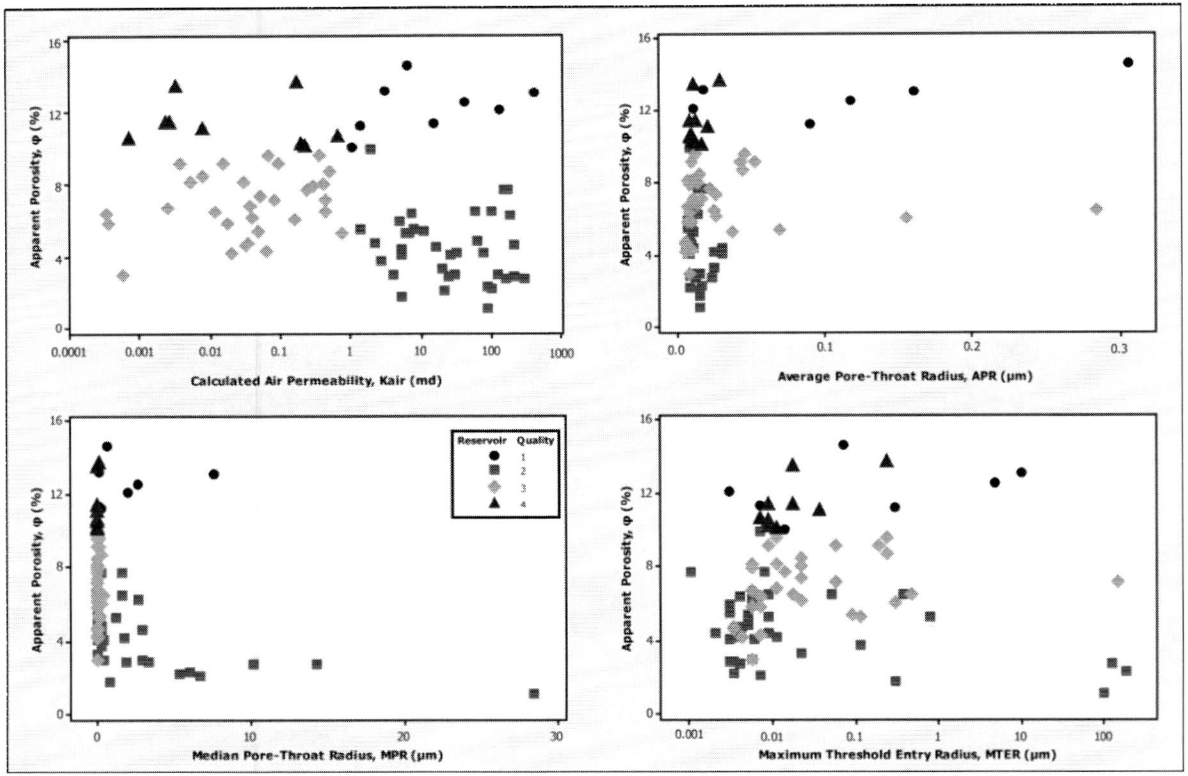

Figure 9. Composite plot of reservoir quality vs. average pore-throat radius, APR (μm), apparent porosity, ϕ (%), median pore-throat radius, MPR (μm), and R35 (μm) values. 1 and 2 refers to reservoir rocks; 3 and 4 refers to non-reservoir rocks. Details are given in text.

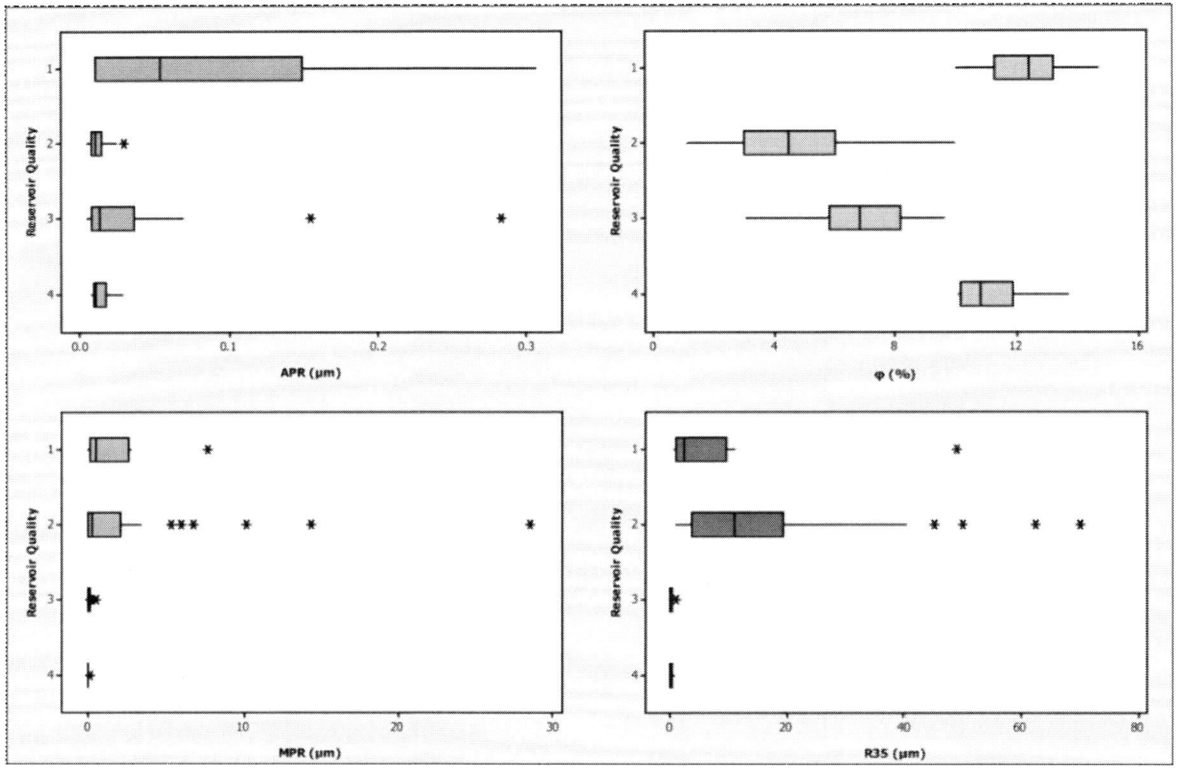

Figure 10. Bloxplots of average pore-throat radius, APR (μm), apparent porosity, ϕ (%), median pore-throat radius, MPR (μm), and R35 (μm) values. 1 and 2 refers to reservoir rocks; 3 and 4 refers to non-reservoir rocks. Details are given in text.

APR exhibits the greatest variation for samples in window 1 (the best possible reservoir rocks), followed by the rock belonging to window 3 (with two outliers). The distinction between reservoir and non-reservoir rocks is not obvious in this case, probably because this parameter is constrained by pore-throat properties (total specific intrusion volume and total specific pore area) irrelevant to reservoir quality.

The boxplots of Φ indicate that rocks in window 1 have, in average, the highest porosities, followed by those belonging to windows 4, 3, and 2, respectively. As for APR distributions, porosity is irrelevant in distinguishing between reservoir and non-reservoir rocks because other parameters (e.g., permeability, R35) play a more important role.

MPR values for reservoir rocks (samples belonging to windows 1 and 2 in Figure 5) exhibit a larger value interval than rocks form windows 3 and 4 (although the median values are comparable for all four windows).

Finally, R35 values display a similar pattern like MPR values, i.e., samples belonging to windows 1 and 2 (reservoir rocks) have a larger value interval, while samples from windows 3 and 4 (non-reservoir rocks) display vary narrow value intervals and median values close to 0 mm. We believe that MPR and R35 are largely controlled by the pore-throat sorting, pore-throat types, and permeability, which are clearly distinct in reservoirs vs. non-reservoir rocks.

We interpret these results as indicating a link between reservoir quality and the distribution of the analyzed petrophysical parameters (Kair, MPR, R35, PTS, and PT).

CONCLUSION

1. Petrophysical analysis is important for understanding pore system geometry and in exploring for and exploiting hydrocarbon reservoirs. MIP measurements and capillary-pressure curves were used to determine several petrophysical characteristics: mean pore-throat radius (MPR), average pore-throat radius (APR), apparent porosity (Φ), calculated air permeability (Kair), port type (PT), pore-throat size distribution (PTD), pore-throat sorting (PTS), maximum-threshold entry radius (MTER), pore-throat radius at 35% mercury saturation (R35), and reservoir quality. Based on the relationships among these petrophysical parameters, we petrophysically characterized the reservoir and non-reservoir rocks of the Anadarko Basin, Oklahoma.
2. MIP technique was used to analyze 91 core samples taken from the reservoir and non-reservoir rocks of the Anadarko Basin. The samples are sandstones (20), shales (24), interbedded shales and siltstones (11), siltstones (17), grainstones (9) and lime mud-stones (10), ranging in age from Devonian to Permian and in depth from 404 m to 4,975 m. The petrophysical parameters used in this study are listed above. The quality of reservoir rocks was estimated using a permeability/porosity/port type plot and separating it into four areas (windows) with various reservoir qualities. Finally, statistical analysis was used to assess correlations between various petrophysical parameters.
3. APR, MTER and Φ values are not well suited to discriminate between reservoir and non reservoir rocks (their value intervals mostly overlap). We surmise that this situation is due to heterogeneous lithology.

4. MPR, R35, and Kair display distinct value intervals for reservoir and non-reservoir rocks. Our interpretation is that the port types mainly control those value intervals, with mostly mega and macro port types for reservoir rocks and mostly micro and nano port types for non-reservoir rocks.
5. Reservoir rocks display mostly (36 out of 50 samples) bimodal pore-throat distributions, while all non-reservoir rocks (41 samples) have unimodal distribution. PT and PTS are considered responsible for this situation.
6. PTS is dominated by PS and MS for reservoir rocks and by WS and MS for non-reservoir rocks. The link between PTS and lithology is clearly exposed in Figure 7. Most samples that have WS PTS are sandstones (44%); most samples that exhibit MS PTS are siltstones (32%), whereas most samples with PS PTS are shales (33%). The distributions are significantly different ($\chi^2 = 23.606$, DF = 10, p-value = 0.009).
7. Two relationships (Φ vs. APR and Φ vs. MTER, both for reservoir rocks) have statistical significant correlation with p-value <0.05. None of statistical correlations has a statistical significance for non-reservoir rocks. We assume that lithological heterogeneity, depositional environments, lateral and vertical change in the formation's mineral composition, with inherent grain fabric, may be responsible for such situations. The different port types (mostly mega and macro sizes for reservoir rocks vs. mostly micro and nano sizes for non-reservoir rocks) could be another explanation.
8. Although modest, the data collected during this research represent the beginning of a petrophysical database for reservoir and non-reservoir rocks of the Anadarko Basin, OK.
9. In this study we illustrated the various links between petrophysical characteristics and distribution of reservoir and non-reservoir rocks in the Anadarko Basin, Oklahoma.
10. This study produced (1) a classification of reservoir vs. non-reservoir rocks based upon their lithology and various petrophysical parameters; (2) identified criteria for classifying reservoir and non-reservoir rocks based on capillary-pressure curves; and (3) established statistical relationships between lithology and petrophysical characteristics of the reservoir and non-reservoir rocks in the Anadarko Basin, OK.

ACKNOWLEDGMENTS

Oklahoma Geological Survey Core and Sample Library in Norman, OK, provided us the core samples used in this research.

REFERENCES

Al-Hinai, S., Fisher, Q. J., Al-Busafi, B., Guise, P. & Grattoni, C. A. (2008). Laboratory measurements of the relative permeability of cataclastic fault rocks: An important consideration for production simulation modeling, *Marine and Petroleum Geology*, v. 25, 473 – 485.

Al-Shaieb, Z., Puckette, J. O., Abdalla, A. A. & Ely, P. B. (1994a). Three levels of compartmen-tation within the overpressured interval of the Anadarko Basin, in P. J. Ortoleva (ed.), Ba-sin Compartments and Seals, edited by, *AAPG Memoir*, v. *61*, 69 – 83.

Al-Shaieb, Z., Puckette, J. O., Abdalla, A. A. & Ely, P. B. (1994b). Megacompartment complex in the Anadarko Basin: a completely sealed overpressured phenomenon, in P. J. Ortoleva (ed.), Basin Compartments and Seals, *AAPG Memoir*, v. *61*, 55 – 68.

Al-Shaieb, Z., Puckette, J., Ely, P. & Tigert, V. (1992). Pressure compartments and seals in the Anadarko Basin, in K. S. Johnson and B. J. Cardott (eds), Source Rocks in the Southern Mid-continent, 1990 Symposium, *Oklahoma Geological Survey Circular*, *93*, 210 – 228.

Aplin, A. C., Fleet, A. J. & Macquaker, J. H. S. (eds.) (1999). *Muds and Mudstones: Physical and Fluid-Flow Properties*, Geological Society Special Publication no. *158*, 190.

Boyd, D. T. (2005). Oklahoma Oil and Gas Production: Its Components and Long-Term Outlook, *Oklahoma Geology Notes*, v. *65*, no.1, 4 – 23.

Coalson, E. B., Hartmann, D. J. & Thomas, J. B. (1985). Productive characteristics of common reservoir porosity types, *S. Tex. Geol. Soc. Bull.*, v. *25*, no. 6, 35 - 51.

Cranganu, C. (2004). Capillary sealing in the Anadarko Basin, *Northeastern Geology & Environmental Sciences*, v. *26*, nos. 1 & 2, 35 - 42.

Cranganu, C. & Villa, M. A. (2006). *Capillary Sealing as an Overpressure Mechanism in the Anadarko Basin*, http://www.searchanddiscovery.net/documents/2006/06019Cranganu/index.htm

Deming, D., Cranganu, C. & Lee, Y. (2002). Self-sealing in sedimentary basins, *J. of Geophys. Res.*, vol. *107*, no. B12, doi: 10.1029/2001JB000504.

Hartmann, D. J. & Beaumont, E. A. (1999). Predicting reservoir system quality and performance, in E. A. Beaumont and N. H. Foster, eds., Exploring for oil and gas traps, *AAPG*, Tulsa, OK, ch. 9.

Hartmann, D. J. & Coalson, E. B. (1990). Evaluation of the Morrow Sandstone in the Sorrento field, Cheyenne Company, Colorado, *Rocky Mountain Association of Geologists*, 91 – 100.

Hill, G. W. & Clark, R. H. (1980). The Anadarko basin – a regional petroleum accumulation – a model for future exploration and development, *Shale Shaker*, v. *31*, 36 – 49.

Jennings, J. B. (1987). Capillary pressure techniques: Application to exploration and development geology, *AAPG Bull.*, v. *71*, 1196 – 1209.

Johnson, K. S. (1989). Geologic evolution of the Anadarko basin, *Oklahoma Geological Survey Circular*, *90*, 3 – 12.

Keighin, C. W. (1997). Physical properties of clastic reservoir rocks in the Uinta, Wind River, and Anadarko basins, as determined by mercury-injection porosimetry, *U. S. Geol. Surv. Bull.*, 2146-G, 73 – 78.

Krushin, J. T. (1997). Seal Capacity of Nonsmectite Shale, in R. C. Surdam, ed., Seals, traps, and the petroleum system, *AAPG Memoir*, *67*, 31 – 47.

Lay, M. (ed.) (2001). Annual review of oil and gas production by fields in the Unites States and Canada, *International Oil Scout Association*, v. *70*, 173 - 221.

Martin, A. J., Solomon, S. T. & Hartmann, D. J. (1997). Characterization of petrophysical flow units in carbonate reservoirs, *AAPB Bull.*, v. *81*, 734 – 759.

Newsham, K. E., Rushing, J. A., Lasswell, P. M., Cox, J. C. & Blasingame, T. A. (2004). *A comparative study of laboratory techniques for measuring capillary pressures in tight gas sands*, SPE 89866. In: SPE Annual Technical Conference and Exhibition, Houston, TX.

Purcell, W. R. (1949). Capillary pressures – Their measurements using mercury and the calculation of permeability therefrom, *American Institute of Mining Metallurgical and Petroleum Engineers Petroleum Transactions*, v. *186*, 39 - 48.

Tanguay, L. H. & Friedman, G. M. (2001). Petrophysical characteristics and facies of carbonate reservoirs: The Red River Formation (Ordovician), Williston basin, *AAPG Bull.*, v. *85*, no. 2, 491 - 523.

Villa, M. A. (2005). Petrophysical characterization of reservoir and non-reservoir formations of the Anadarko Basin, based on mercury intrusion porosimeter measurements, MA Thesis, Brooklyn College, 188.

Washburn, E. W. (1921). Note on a method of determining the distribution of pore sizes in a porous material, *Proceedings of the National Academy of Science*, v. *7*, 115 -116.

Webb, P. A. (2001). An Introduction To The Physical Characterization of Materials by Mercury Intrusion Porosimetry with Emphasis On Reduction And Presentation of Experimental Data, *Micromeritics Instrument Corp.*, Norcross, Georgia, 22.

Winland, H. D. (1972). Oil accumulation in response to pore size changes, Weyburn field, Saskatchewan, Amoco Production Company Report F72-G-5, 20. (unpublished).

Winland, H. D. (1976). Evaluation of gas slippage and pore aperture size in carbonate and sandstone reservoirs, Amoco Production Company Report F76-G-5, 25 (unpublished).

Yang, Y. & Aplin, A. C. (2007). Permeability and petrophysical properties of 30 natural mudstones, *Journal of Geophysical Research*, vol. *112*, B03206, doi: 10.1029/2005JB004243.

In: Natural Gas and Petroleum
Editor: Constantin Cranganu

ISBN: 978-1-62808-535-8

Chapter 4

CURRENT SITUATION OF NATURAL GAS AND PETROLEUM

L. C. Castañeda* and C. Marín
Instituto Mexicano del Petróleo, México, D.F., México

ABSTRACT

Natural gas (NG) and petroleum, the most important sources of energy around the world are discussed in this chapter. Relevant information regarding the NG and petroleum reserves, production and consumption and NG and oil prices is reported. The stock of proved reserves of natural gas distribution, in which Russia is the biggest holder, is included in this chapter. In case of petroleum, the evolution of proved reserves on each region is analyzed where most of the world oil reserves are in the Middle East. The distribution of proved reserves by region was analyzed to determine the ten top world reserves of Natural gas and petroleum and the prediction on what is expected in the coming years. As with other fossil fuels, burning natural gas produces carbon dioxide, which is the most important greenhouse gas. Since many scientists agree that high levels of greenhouse gas are changing the global climate, an analysis of the potential effect of increasing levels of carbon dioxide and other greenhouse gases in the earth's atmosphere is also presented.

Keywords: Natural Gas, Petroleum, Heavy oil

NOMENCLATURE

Coalbed methane: Coalbed gas or coal mine methane (CMM) is a form of natural gas extracted from coal beds. CBM consists of over 90% methane, which is normally stable, meaning that the gas can be fed directly into the natural gas network or a gas engine.

Natural gas resource base: An estimate of the amount of natural gas available, based on the combination of proved reserves, and those additional volumes that have not yet been

* Eje Central Lázaro Cárdenas Norte 152, 07730 México, D.F., México. E-mail: lcastane@imp.mx.

discovered, but are estimated to be "discoverable" given current technology and economics.

Proved amount in place: The volume originally occurring in known natural reservoirs that has been carefully measured and assessed as exploitable under present and expected local economic conditions with existing available technology.

Proved recoverable reserves: The volume within the proved amount in place that can be recovered in the future under present and expected local economic conditions with existing available technology.

Proved Resources: Part of the resource base that includes the working inventory of natural gas; volumes that have already been discovered and are readily available for production and delivery.

Shale gas: Is natural gas formed from being trapped within shale formations.

Raw natural gas: Impure natural gas as delivered from the well and before processing (refining).

Reserves: The amount of a resource available for recovery and/or production; the recoverable amount is usually tied to economic aspects of production.

Tight gas: Unconventional natural gas which is difficult to access because of the nature of the rock and sand surrounding the deposit. This gas is much more difficult to extract than natural gas from other sources, therefore, hydraulic fracturing and directional drilling is necessary to produce the well.

Petroleum Reserves. Those quantities of petroleum which are anticipated to be commercially recovered from known accumulations from a given date forward. The relative degree of uncertainty may place reserves in two principal classifications: proved or unproved.

Proved Reserves. Proved reserves are those quantities of petroleum which, by analysis of geological and engineering data, can be estimated with reasonable certainty to be commercially recoverable, from a given date forward, from known reservoirs and under current economic conditions, operating methods, and government regulations. Proved reserves can be categorized as developed or undeveloped.

Unproved Reserves. Unproved reserves are based on geologic and/or engineering data similar to that used in estimates of proved reserves; but technical, contractual, economic, or regulatory uncertainties preclude such reserves being classified as proved.

Probable Reserves. Probable reserves are those unproved reserves which analysis of geological and engineering data suggests are more likely than not to be recoverable. In this context, when probabilistic methods are used, there should be at least a 50% probability that the quantities actually recovered will equal or exceed the sum of estimated proved plus probable reserves.

Possible Reserves. Possible reserves are those unproved reserves which analysis of geological and engineering data suggests are less likely to be recoverable than probable reserves. In this context, when probabilistic methods are used, there should be at least a 10% probability that the quantities actually recovered will equal or exceed the sum of estimated proved plus probable plus possible reserves.

INTRODUCTION

1. Natural Gas

Natural gas is a really important source of energy for reducing pollution and preserving a clean and healthy environment. The use of natural gas offers several environmental benefits over other sources of energy, specifically, other fossil fuels in addition to being an abundant and secure source of energy in several regions such as the United States, Asia and Europe.

Natural gas burns more cleanly than other fossil fuels. It has fewer emissions of sulfur, carbon, and nitrogen than coal or oil, and it has almost no ash particles left after burning (Speight).

The primary component of natural gas is methane (CH_4), and it also contains gaseous hydrocarbons such as ethane (C_2H_6), propane (C_3H_8), and butane (C_4H_{10}), as well as other non-hydrocarbon gases. In addition, raw natural gas contains water vapor, hydrogen sulfide (H_2S), carbon dioxide, helium, nitrogen, and other compounds (EIA 1999a, Gallagher). Table 1 shows a composition of the natural gas.

The natural gas used by consumers is composed almost completely of methane. Raw natural gas comes from three types of wells: oil wells, gas wells, and condensate wells. Natural gas that comes from oil wells is typically known as 'associated gas'. This gas can exist separate from oil in the formation (free gas), or dissolved in the crude oil (dissolved gas). Natural gas from gas and condensate wells, in which there is little or no crude oil, is termed non associated gas (NAG). Gas wells typically produce raw natural gas by itself, while condensate wells produce free natural gas along with a semi-liquid hydrocarbon condensate (Devold).

Table 1. Gulf of Mexico (GOM) gas composition

Component	Formula	Production Sales Mole %	Inlet to gas processing plant Mole %
Methane	CH_4	88.023	90.00
Ethane	C_2H_6	5.824	4.65
Propane	C_3H_8	3.292	2.50
Isobutane	C_4H_{10}	0.936	0.50
n-Butane	C_4H_{10}	0.537	0.75
Isopentane	C_5H_{12}	0.249	0.20
n-Pentane	C_5H_{12}	0.236	0.40
n-Hexane	C_6H_{14}	0.149	0.20
n-Heptane	C_7H_{16}	0.189	0.10
n-Octane	C_8H_{18}	0.098	0.05
n-Nonane	C_9H_{20}	0.036	-
Nitrogen	N_2	0.262	0.40
Carbon dioxide	CO_2	0.169	0.25

Natural gas processing consists of separating all of the various hydrocarbons and fluids from the pure natural gas, to produce what is known as dry natural gas (pipeline quality). Major transportation pipelines usually impose restrictions on the makeup of the natural gas that is allowed into the pipeline and measure energy content in kJ/kg (also called calorific value or wobbe index) (Devold).

2. Reserves

Reserves are defined as the amount of a resource available for recovery and/or production in the future, the recoverable amount is usually coupled to economic aspects of production (Speight).

Contrasting the inventory of a manufacturing company, reserves are physically located in deep underground reservoirs that cannot be visually inspected or counted. Hence, estimates based on the evaluation of data that provides evidence of the amount of oil and gas present is usually applied. All reserve estimates involve some degree of uncertainty; however, the estimation of reserves volumes is generally performed by expert individuals who use their experience and qualified professional judgment in the calculation of those volumes. Reserves also reflect the potential value that will have to be shared between companies and the populations represented by their governments (SPE 2007, SPE 2011).

Proved gas reserves are commonly used as a measure of geographical gas distribution in the world, but they are not a straightforward indicator of the productive capacity of a country. Proved reserves are only so classified when there is a high probability they are recoverable. Figure 1 shows a world distribution of proved reserves of natural gas and Figure 2 shows the top ten countries with the highest proved gas reserves.

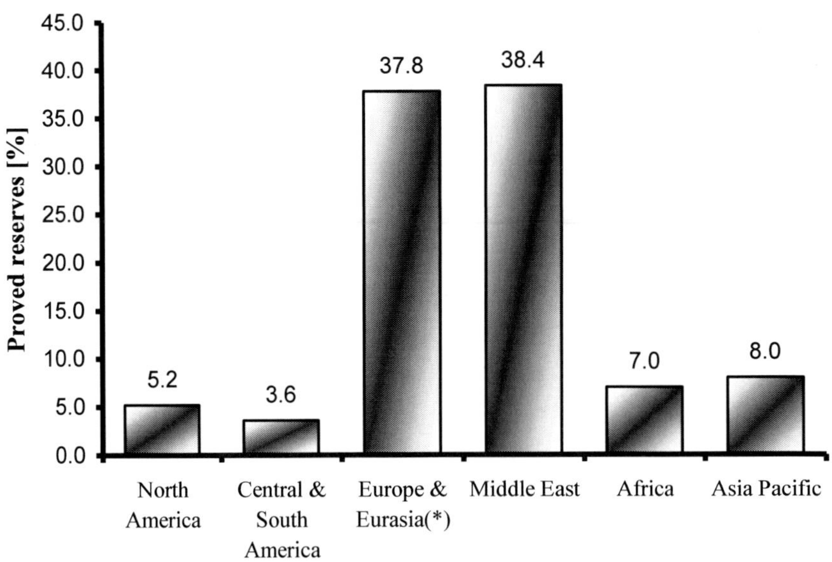

(SPE 2007, British Petroleum 2012)

Figure 1. Geographical distribution of proved gas reserves at the end of 2011. (*) Eurasia includes Russia, Central Asia and Caucasus. 100% = 208.4 Trillion cubic meters.

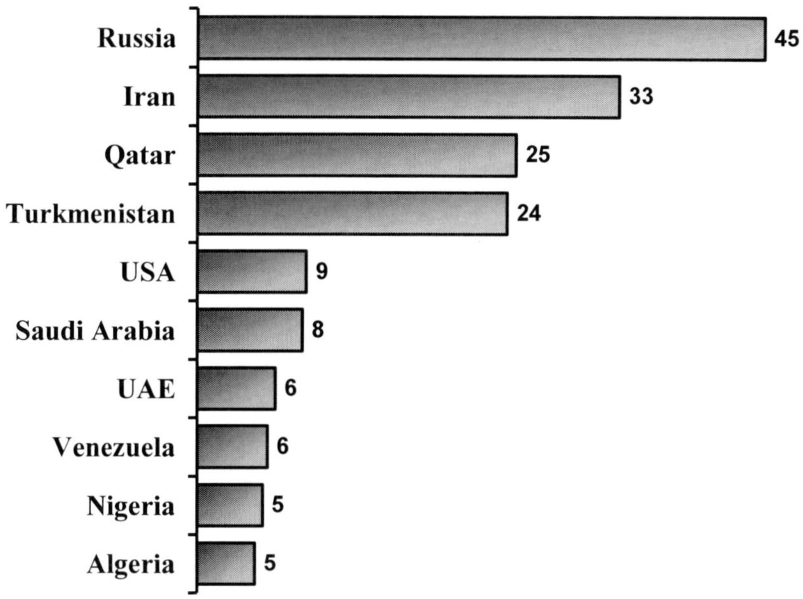

(British Petroleum 2012).

Figure 2. Countries with the highest proved gas reserves at the end of 2011. Trillion Cubic Meters.

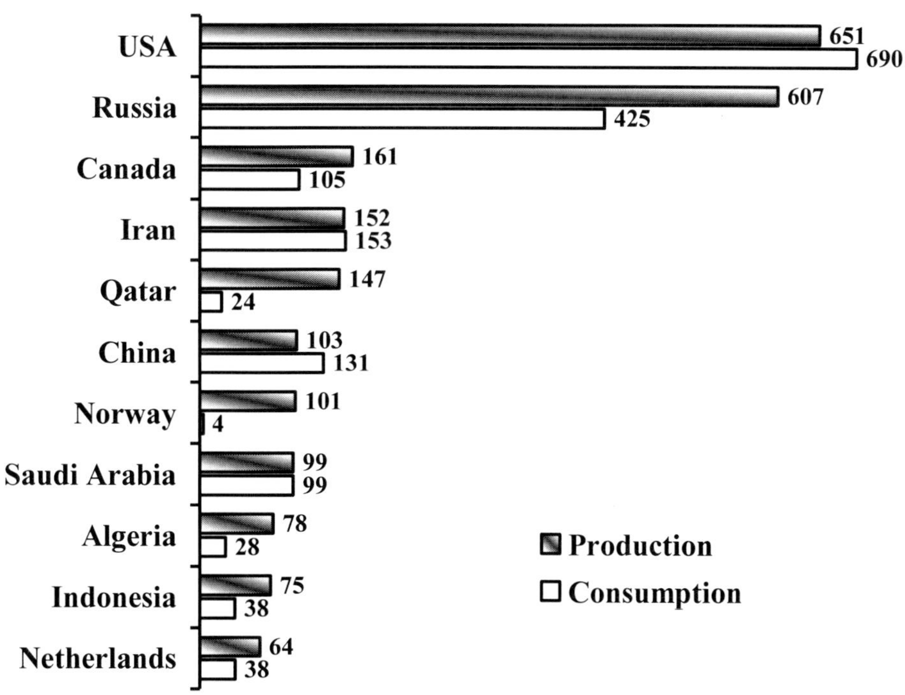

(British Petroleum 2012).

Figure 3. Total production and consumption of Top ten gas producers for 2011.

Although Europe and Eurasia group have similar reserves as Middle East, Eurasia accounted for the highest contribution. The main reason for this increasing is because Turkmenistan has shown a strong increasing with 2.6 MMCM in 2001, 13.4 in 2010, and almost 24 MMCM in 2011 as shown in Figure 1. Other countries such as Russia have kept almost constant its proved reserves, i.e. 42.4, 44.4 and 44.6 for the same years.

2.1. Production and Consumption

United States maintains the world's largest Natural gas production capacity, to be around 650 billion cubic meters (BCM) in 2011. The highest consumption is also observed in USA. Figure 3 shows the total production and consumption of natural gas of the ten top countries (British Petroleum 2012). Last decade was dominated by Russian Federation with an average production of 569 Billion cubic meters, when the average production in USA accounted for 550 BCM. This situation of Russian domination was not observed during all last ten years, because from 2009 to 2011 the first position is held by US with an average production of 613 BCM versus 575 BCM produced by Russia.

In Middle East, Iran and Saudi Arabia maintain equilibrium because the production and consumption are essentially the same; other countries such as Qatar and Norway are basically net exporters.

Table 2. Comparison of natural gas projections, 2015, 2025, and 2035 (trillion cubic feet)

Projection	2010	AEO2012 Reference case	Other projections					
			IHSGI	EVA	Deloitte	SEER	Exxon Mobil	INFORUM
					2015			
Dry gas production[a]	21.58	23.65	23.81	23.80	24.52	23.66	24.00	24.29
Consumption	24.13	25.39	25.52	26.60	24.07[b]	26.05	25.00[c]	23.61[b]
Henry Hub Spot Price[d]	4.39	4.29	4.75	4.07	4.25	4.28	NR	NR
					2025			
Dry gas production[a]	21.58	26.28	27.23	26.70	27.32	25.88	27.00	27.57
Consumption	24.13	25.53	29.39	29.00	26.36[b]	27.10	29.00[c]	23.43[b]
Henry Hub Spot Price[d]	4.39	5.63	4.82	6.47	5.80	6.29	NR	NR
					2035			
Dry gas production[a]	21.58	27.93	31.35	NR	27.87	27.00	26.00	30.71
Consumption	24.13	26.63	33.54	NR	27.30[b]	27.24	29.00[c]	24.66[b]
Henry Hub Spot Price[d]	4.39	7.37	5.13	7.26	6.63	7.70	NR	NR

NR = not reported.

[a] Does not include supplemental fuels.

[b] Does not includes lease, plant, and pipeline fuel and fuel consumed in natural gas vehicles.

[c] Does not includes lease, plant, and pipeline fuel.

[d] 2010 dollars per million Btu.

2.2. Projections

Natural gas is the fossil fuel with the fastest growing in the world. The projections of natural gas production, consumption, and prices vary considerably due to differences in assumptions (Table 2). One assumption is that current laws and regulations remain unchanged during the projection period, whereas the other projections may include anticipated policy developments over the next 25 years. In particular, the AEO2012 reference case does not assume changes in CO_2 emissions policies (EIA 2012).

Each of the projections shows an increase in overall natural gas consumption from 2010 to 2035, with the IHS Global Insight (IHSGI) projection showing the largest increase, 39%. The ExxonMobil projection includes an increase of around 20%. The Energy Ventures Analysis (EVA) projection shows an increase of 26% from 2010 to 2030 (EVA does not extend to 2035). Total natural gas consumption in the AEO2012, Deloitte, and Strategic Energy and Economic Research, Inc. (SEER) projections increases from 2010 to 2035, with total natural gas consumption growing from 4 to 31%. IHSGI shows the largest increase and Interindustry Forecasting Project at the University of Maryland (INFORUM) the smallest. The IHSGI projection for total natural gas consumption in 2035 is 36% higher than the INFORUM projection. In the AEO2012 Reference case, total natural gas consumption grows by 5% from 2015 to 2035.

The growth in consumption of natural gas occurs in every region and is most concentrated in non-OECD countries, where demand increases nearly three times as fast as in The Organisation for Economic Co-operation and Development (OECD) countries (Figure 4). Increases in production in the non-OECD regions more than meet their projected consumption growth, and as a result non-OECD exports to OECD countries grow through 2035. Non-OECD producers account for more than 81% of the total growth in world natural gas production from 2008 to 2035 (EIA 2011).

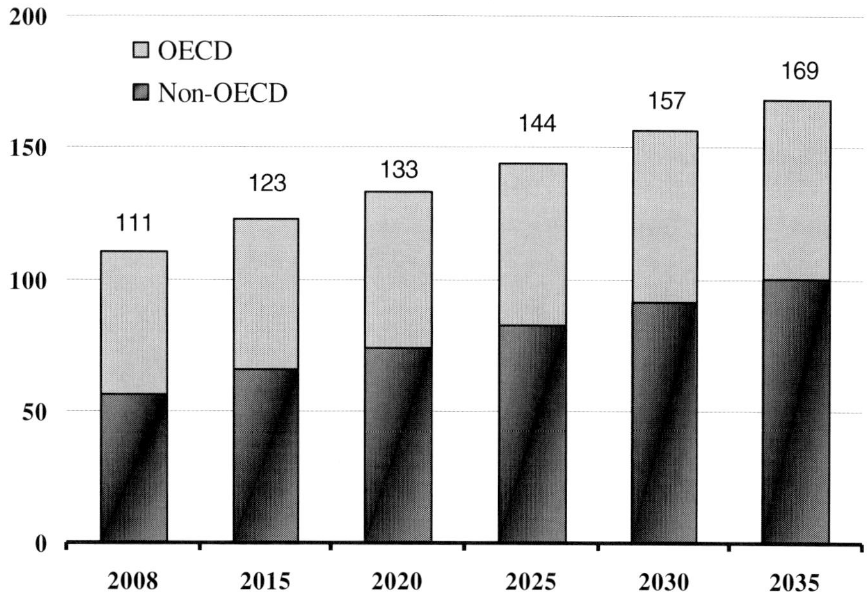

(EIA 2012).

Figure 4. World natural gas consumption, 2008-2035. Trillion cubic feet.

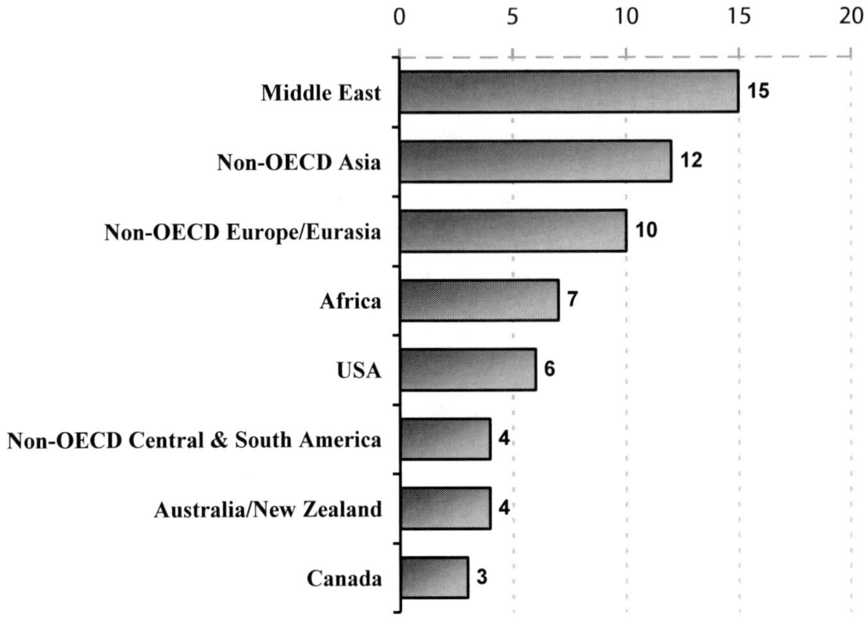

Figure 5. Change in world natural gas production by region 2008-2035 (trillion cubic feet).

The global recession of 2008-2009 resulted in a decreasing of nearly 4% in natural gas demand in 2009. As the recession retreated and economic growth restarted, natural gas demand reached an estimated 113.1 trillion cubic feet (TCF) in 2010, greater than annual consumption levels before the economic decline (EIA 2011, Cedigaz). Natural gas continues to be favored as an environmentally attractive fuel relative to other hydrocarbon fuels. Natural gas consumption grows from 110.7 TCF in 2008 to 168.7 TCF in 2035.

Growth in natural gas consumption is particularly strong in non-OECD countries, where economic development leads to increased demand over the projection period. Consumption in non-OECD countries increases by an average of 2.2% per year through 2035, nearly three times as fast as the 0.8% annual growth rate projected for natural gas demand in the OECD countries. As a result, non-OECD countries account for 76% of the total world increment in natural gas consumption, as the non-OECD share of world natural gas use increases from 51% in 2008 to 59% in 2035.

Natural gas continues to be the fuel of choice in many regions of the world in the electric power, and industrial sectors, in part because of its lower carbon intensity compared with coal and oil, and also because of its significant price discount relative to oil in many world regions. Total world natural gas consumption for industrial uses increases by an average of 1.7% per year through 2035, and consumption in the electric power sector grows by 2.0% per year. The industrial and electric power sectors together account for 87% of the total projected increase in natural gas consumption.

Contributing to the strong competitive position of natural gas among other energy sources is a strong growth situation for reserves and supplies. Significant changes in natural gas supplies and global markets continue with the expansion of liquefied natural gas (LNG) production capacity, even as new drilling techniques and other efficiencies have made production from many shale basins economical worldwide (EIA 2011).

The largest production increases from 2008 to 2035 (Figure 5) are projected for the Middle East (15.3 trillion cubic feet) and non-OECD Asia (11.8 trillion cubic feet). Iran and Qatar increase natural gas production by a combined 10.7 trillion cubic feet, or nearly one-fifth of the total increment in world gas production. A significant share of the increase is expected to come from a single offshore field, which is called North Field on the Qatari side and South Pars on the Iranian side.

Although the extent of the world's unconventional natural gas resource base (which consists of tight gas, shale gas, and coalbed methane) has not yet been assessed fully, the IEO2011 reference case projects a substantial increase in those supplies, especially in the United States and also in Canada and China. In the United States, one of the keys to increasing natural gas production has been advances in the application of horizontal drilling and hydraulic fracturing technologies, which have made it possible to develop the vast shale gas resources and contributed to a near doubling of total U.S. technically recoverable natural gas resource estimates over the past decade. In the Reference case, shale gas accounts for 47% of U.S. natural gas production in 2035. Unconventional resources are even more important for the future of domestic natural gas supplies in Canada and China, where they account for 51% and 72% of total domestic production, respectively, in 2035 in the reference case (EIA 2011, EIA 2006).

World natural gas liquefaction (LNG) capacity nearly doubles, from about 8 trillion cubic feet in 2009 to 15 trillion cubic feet in 2035. Most of the increase in liquefaction capacity is in the Middle East and Australia, where a multitude of new liquefaction projects are expected to be developed, many of which will be operating within the next decade. The liquefaction capacity is expected to stay high over the entire projection. Given the capital-intensive nature of liquefaction projects, long-term contracts requiring the purchase of high volumes often are used to ensure high utilization rates and acceptable returns on investments (McCoy, Clarke).

2.3. Natural Gas and the Environment

Natural gas is a mainly important source of energy for reducing pollution and maintaining a clean and healthy environment. In addition to being an abundant and secure source of energy in many countries such as the United States and Germany, the use of natural gas also offers several environmental benefits over other sources of energy, particularly other fossil fuels. Natural gas burns more cleanly than other fossil fuels. It has fewer emissions of sulfur, carbon, and nitrogen than coal or oil, and it has almost no ash particles left after burning. Being a clean fuel is one reason that natural gas is preferred, especially for electricity generation, it has maintained a continuous growth and is expected to grow even more in the future. As with other fossil fuels, burning natural gas produces carbon dioxide (CO_2), which is the most important greenhouse gas. Many scientists believe that increasing levels of carbon dioxide and other greenhouse gases in the earth's atmosphere are changing the global climate (Speight).

As with other fuels, natural gas also affects the environment when it is produced, stored, and transported. Because of methane (another greenhouse gas) is the major component of natural gas, small amounts of methane can sometimes leak into the atmosphere from wells, storage tanks, and pipelines. Exploring and drilling for natural gas have some impact on land and marine habitats. However new technologies have greatly reduced the number and size of areas disturbed by drilling. Natural gas pipelines and storage facilities have a very good safety record, which is important because when natural gas leaks a high risk of explosions is present.

On a relative basis, natural gas is the cleanest of all the fossil fuels. Composed primarily of methane, the main products of the combustion of natural gas are carbon dioxide and water vapor. Coal and petroleum are composed of much more complex molecules, with a higher carbon ratio and higher nitrogen and sulfur contents. Thus, when combusted, coal and oil release higher levels of harmful emissions, including a higher ratio of carbon, nitrogen oxides (NO_x), and sulfur dioxide (SO_2). Coal and fuel oil also release ash particles into the environment, substances that do not burn but instead are carried into the atmosphere and contribute to pollution. The combustion of natural gas, on the other hand, releases very small amounts of sulfur dioxide and nitrogen oxides, virtually no ash or particulate matter, and lower levels of CO_2, carbon monoxide (CO), and other reactive hydrocarbons (Table 3). The use of fossil fuels for energy contributes to a number of environmental problems. Natural gas, as the cleanest of the fossil fuels, can be used in many ways to help reduce the emissions of pollutants into the atmosphere. Burning natural gas instead of other fossil fuels emits fewer harmful pollutants into the atmosphere, and an increased reliance on natural gas can potentially reduce the emission of many of these harmful pollutants (EIA 2006). Pollutants emitted in the United States, particularly from the combustion of fossil fuels, have led to the development of many pressing environmental problems. Natural gas, emitting fewer harmful chemicals into the atmosphere than other fossil fuels, can help to mitigate some of these environmental issues.

A recent report of International Energy Agency (IEA 2012) indicates that the averaged World CO_2 emissions from electricity generation using coal/peat are 971 grams CO_2/kilowatt hour, for 2008-2010 years. The corresponding CO_2 emissions from electricity generation using oil and using natural gas are 779 and 450 grams CO_2/kilowatt hour, respectively.

The top 10 countries that emit the highest carbon dioxide emissions according with the International Energy Agency are shown in Figure 6. Nearly two-thirds of global emissions for 2010 are originated from just ten countries, with the contribution of China (23.8%) and the United States (17.7%) far exceeding those of all others. Combined, these two countries alone produced 12.6 Gt CO_2, 41.5% of world CO_2 emissions.

China owns the number one position for the world's most polluting country as it emitted 7,217 million tonnes of greenhouse in 2010. This is explained because enormous amounts of goods are manufactured in China. China also has the largest population in the world so it consumes huge amounts of fossil fuels for transportation, cities, workplaces and food production (AFOP 2012).

Table 3. Fossil Fuel Emission Levels-lb/10^8 Btu of Energy Input

Pollutant	Natural Gas	Petroleum	Coal
Carbon dioxide	117,000	164,000	208,000
Carbon monoxide	40	33	208
Nitrogen oxides	92	448	457
Sulfur dioxide	1	1,122	2,591
Particulates	7	84	2,744
Mercury	0.000	0.007	0.016

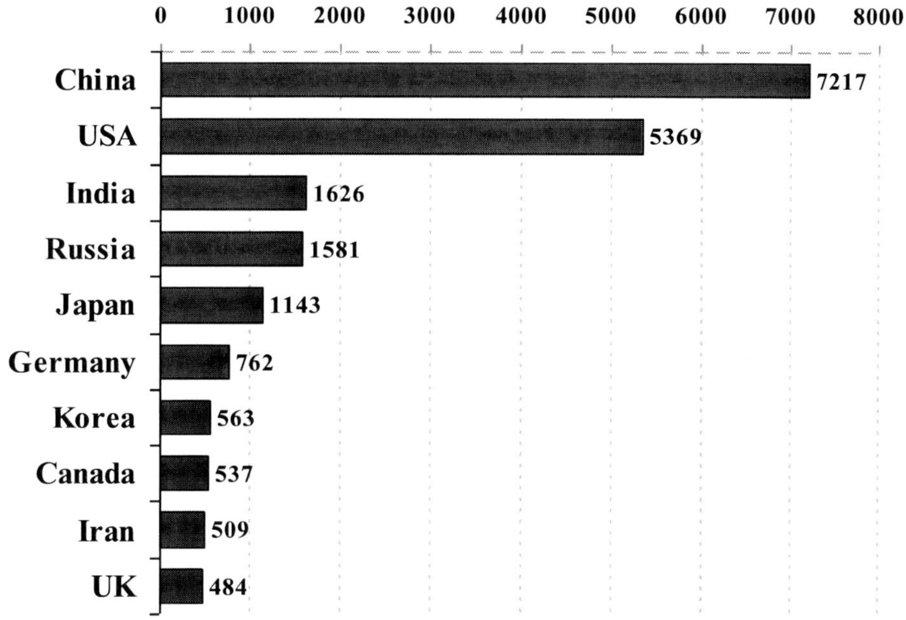

Figure 6. Top ten emitting countries in 2010 account for 65% of the world CO_2 emissions. World total: 30.3 $GtCO_2$ (Gt: Millions of metric tons).

Prior to 2007 the United States was the biggest polluter country; it still comes not too far behind China producing 5,369 million tonnes of greenhouse gases. The USA consumes 19,180,000 barrels of oil per day, because the operation of one of the largest meat industries in the world and besides other largest industries. Its big population is also consuming oil for transportation, food production, their homes utilities and their workplaces.

India is the second highest populated country in the world. As a result, the country has a very high carbon output of 1,293 million tonnes. The country is becoming heavily industrialized and is undergoing significant growth in production and exportation.

Russia has a smaller population compared with the US and China but still produces a significant amount of CO_2, 1,581 million tonnes per year. Russia pronounces it is ready to diminish greenhouse gas emissions by 25% when other countries agree to do the same.

Japan produced 1,143 million tonnes of CO_2 in the same year. Similar to Russia, Japan plans to cut its CO_2 emissions by 25% when it sees other countries actions aimed the same route. The main contributors to the pollution are the large areas of urbanization, industrialization and transportation. The Japanese Environment Agency (JEA) has taken steps to reduce emissions by adding regulations for automobile gas and tightening controls on nitrogen dioxide emission.

Germany produced considerably less CO_2 than Japan with 762 million tonnes. Germany has ambitious goals to cut CO_2 emission by 3% a year in order to reach United Nations targets. The German government also encourages citizens to cut their electricity use in an attempt to reach a 2020 goal of a 13% reduction in electricity consumption.

South Korea comes in 7th with 563 million tonnes of carbon dioxide. Since 1960, there has been heavy industrialization and a constant demand for the latest technologies. These technologies are often produced in the country, so there is a large manufacturing industry.

Factory and transportation emissions are the main contributors to sulfur-oxides being emitted into the atmosphere.

Canada is the only country on the list that may appear as a surprise. The country produced 537 million tonnes of carbon dioxide. Stories of sustainable developments and high standard of living in Canada would make not suspect of its high pollution levels. Experts place much of the blame on the mining industry, the strong dependence on road transportation and disposal of chemicals contaminating the air and water.

Iran supplies 509 million tonnes of carbon dioxide annually. Iran suffers problems with air and water pollution with the ecology in the Caspian Sea drastically threatened. Household waste, factory chemicals and contaminators are often dumped into the Caspian.

The final country on the list, the U.K creates 484 million tonnes of CO_2 yearly. While air pollution is being combated and energy efficiency is improving, transportation pollution still remains a major problem. Burning fossil fuels like coal and oil, for energy, still remains a large problem for the UK. In contrast, recently Britain opened the largest wind farm in the world and there are now plans for an even bigger sustainable power project.

3. Petroleum

Petroleum is a complex mixture of liquid hydrocarbon occurring naturally within the earth. Although petroleum provides many useful products, the most notable are motor gasoline and heating fuel. Petroleum is mostly used in the transportation sector include fuel for automobiles, trucks, agricultural and industrial machinery, trains, ships, and aircraft. Petroleum is used to heat homes, offices, and factories and is used to grow, process, package, distribute, refrigerate, and cook food. Petroleum is also the source of synthetic fabric for cloths as well as detergents and dry cleaning solvent to clean them. Moreover, petroleum provides a chemical base for cosmetics and pharmaceutical products as well as for many plastic products from toys to building materials (EIA 1999b). Crude oils properties such as API gravity and sulfur content are typically used to classify the quality and price of every crude oil. Figure 7 shows the properties of the most important crude oils around the world.

3.1. Reserves

Reserves are those quantities of petroleum declared to be commercially recoverable by application of development projects to known accumulations under defined conditions. Because all reserve estimates involve uncertainty, the relative degree of uncertainty can be expressed by dividing reserves into two major classifications: proved and unproved. Unproved reserves can further be divided into two subcategories, probable and possible, to specify the relative degree of uncertainty about their existence (SPE 2007). The most commonly accepted definitions of these reserves are based on those approved by the Society of Petroleum (SPE) and the World Petroleum Council (WPC) in 1997 (SPE 1997, Research and Markets).

According to BP Statistical review of world energy, as of June 2012 the world crude oil proved reserves during the last years have slightly increased. Figure 8 shows the evolution of proved reserves on each region where most of the world oil reserves are in the Middle East (British Petroleum 2012, EIA 2011). The remaining regions account for 52% of the world oil reserves where South-Central America has the second place with almost 20% and North America is in the third place with 13% of those reserves.

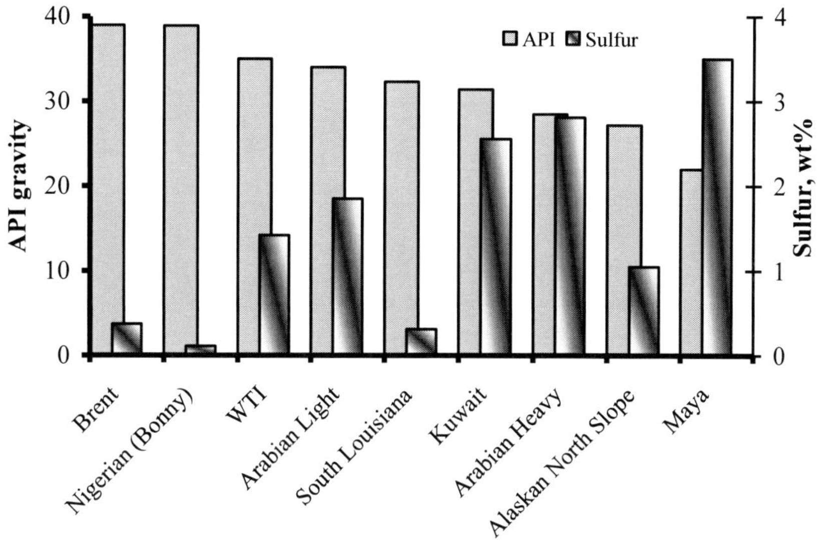

Figure 7. Properties of the most important crudes oils around the world.

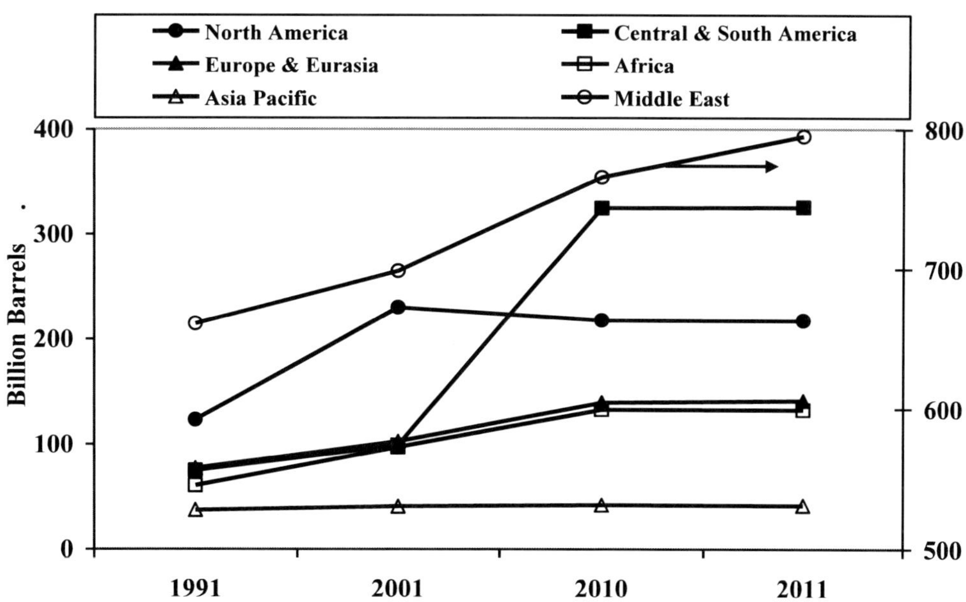

Total Billion Bbl	**1033**	**1267**	**1622**	**1653**

(British Petroleum 2012, EIA 2011).

Figure 8. Evolution of world crude oil proved reserves by region.

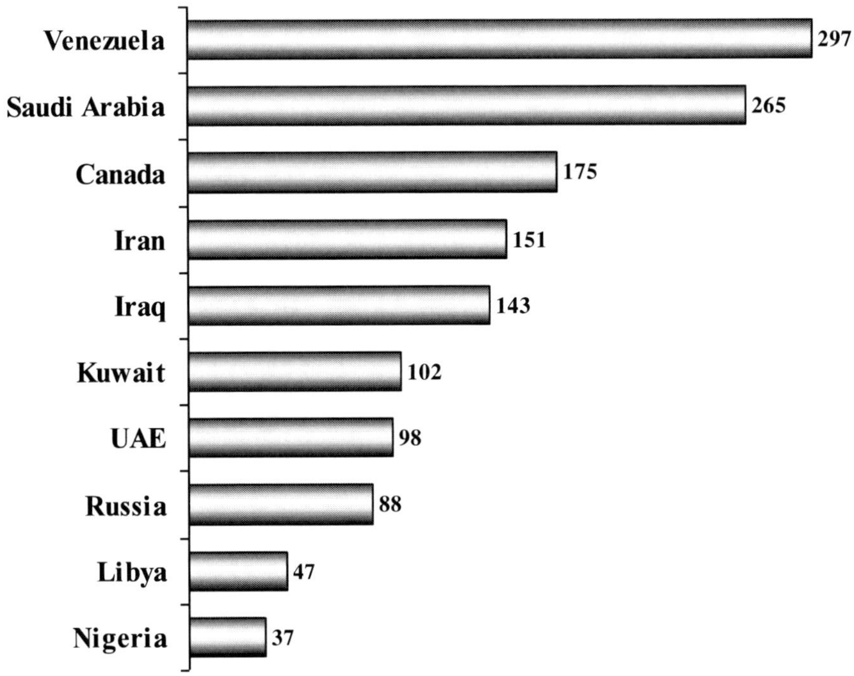

(EIA 2011).

Figure 9. Top ten proved world oil reserves, Billions of barrels (2011).

Saudi Arabia held the highest petroleum reservoirs during last two decades, but since 2010, Venezuela has the top position accounting for 296.5 billion barrels of proven oil reserves, roughly 18% of the world total which economy remains seriously dependent on crude oil. Although Saudi Arabia has around 100 major oil and gas fields (and more than 1,500 wells), over half of its oil reserves are contained in only eight fields.

Canada had 175.2 billion barrels of proven oil reserves at end 2011, third in the list. The bulk of these reserves (over 95%) are oil sands deposits in Alberta, which are more difficult to extract and process than conventional crude oil. Canada is a net exporter of oil where almost all of the exports flow to the United States, and it is consistently the highest supplier of US oil imports (EIA 2010).

Last two years, Iraq, Kuwait, Venezuela, UAE and Russia essentially maintained their proven oil reserves constant. Figure 9 shows the distribution of the top ten world oil reserves.

3.2. Heavy Oil

There are very large heavy oil and extra-heavy oil resources whose extent and locations are well known. The International Energy Agency (IEA) estimates that there are 6 trillion (6·1012) barrels in place worldwide. Most of these resources are currently untapped (Vartivarian).

Heavy oil and bitumen are widely distributed around the globe and have been found on every continent except Antarctica. Discovered original oil in place includes 3,366 billion barrels of conventional heavy oil with API gravities between 10 and 20 degrees. Non-conventional resources include 2,500 billion barrels of bitumen with another billion or so estimated to be undiscovered and 2,000 billion barrels of extra-heavy oil. Most of the heavy crudes that are produced today come from North and South America, with significant quantities being produced in the United States, Canada, Mexico, Venezuela and Brazil (Solari).

The main problems that heavy crude oil presents are: low processing capacity in the refineries, low mobility through the reservoir because of its high viscosity, and difficult and costly transportation from the platform to the ground and to the refineries.

3.3. Production and Consumption

Saudi Arabia maintains the world's largest crude oil production capacity, to be around 11 million bbl in 2011. Figure 10 shows the crude oil production and consumption of countries belonging to the OPEC (EIA 2010).

Over 2 million bbl/d of capacity was added in 2009 with the addition of increments at Khurais, AFK, Shaybah, and Nu'ayyim. Saudi Arabia is the largest oil consuming country in the Middle East. In 2008, Saudi Arabia consumed approximately 2.4 million bbl/d of oil, an increase of 50% since 2000, due to strong economic and industrial growth and subsidized prices (EIA 2010).

3.4. Projections

In the AEO2012 Reference case, the average price for WTI crude oil (in 2010 dollars) increases to $116.91 per barrel in 2015, $132.56 per barrel in 2025, and $144.98 per barrel in 2035 (Table 4). Prices are lower in the EVA and P&G projections, ranging from $82.24 per barrel in 2015 to $102.11 per barrel in 2035 for EVA, and $98.75 in 2015 to $107.37 in 2035 for P&G. INFORUM and Purvin & Gertz (P&G) did not report projections of WTI prices (EIA 2012).

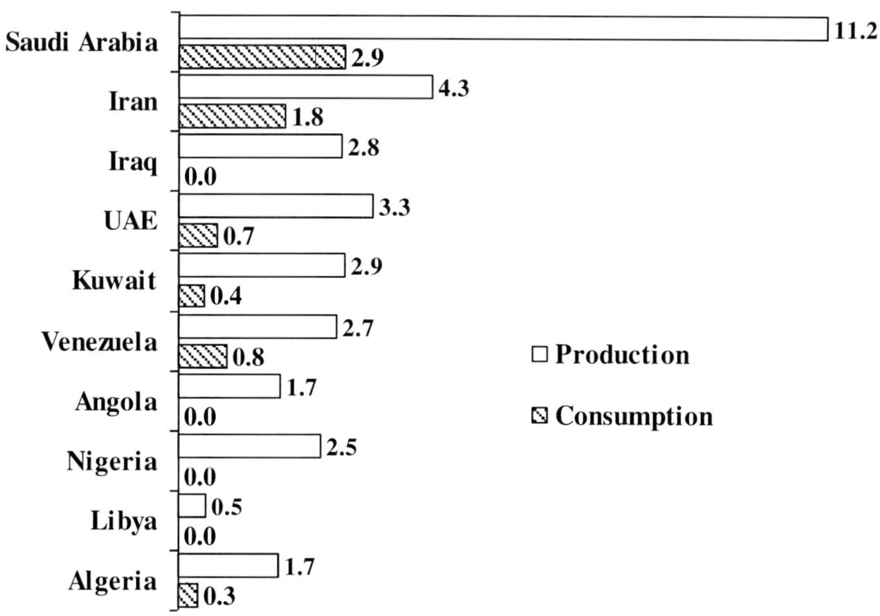

(EIA 2010).

Figure 10. OPEC crude oil production and consumption for 2011, MMBPD.

Table 4. Comparison of liquids projections for 2015, 2025, and 2035

Projection	2010	AEO2012 Reference case	Other projections BP	EVA	INFORUM	P&G
			2015			
Average WTI price[a]	79.39	116.91	--	82.24	--	98.75
Crude oil Production[b]	5.47	6.15	--	6.90	5.43	5.43
Liquids consumption	19.17	19.10	18.26	--	20.04[c]	17.69
			2025			
Average WTI price[a]	79.39	132.56	--	89.07	--	106.47
Crude oil Production[b]	5.47	6.40	--	7.10	5.74	4.26
Liquids consumption	19.17	19.20	17.30	--	20.38[c]	17.39
			2035			
Average WTI price[a]	79.39	144.98	--	102.11	--	107.37
Crude oil Production[b]	5.47	5.99	--	--	5.80	3.23
Liquids consumption	19.17	19.90	--	--	21.31[c]	17.38

[a] 2010 dollars per barrel.

[b] Domestic.

[c] For INFORUM, liquids demand data were converted from quadrillion Btus to barrels at 187.84572 million barrels per quadrillion Btu.

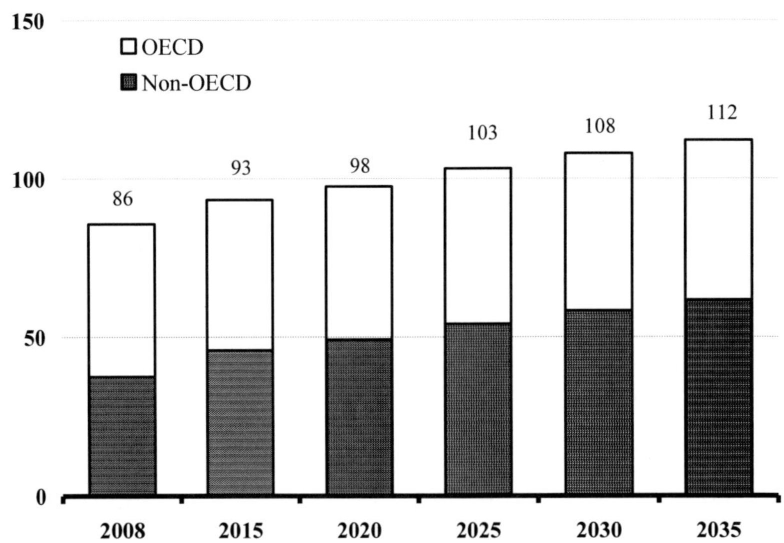

(EIA 2012, EIA 2011).

Figure 11. World liquids consumption, 2008-2035. Million barrels per day.

Domestic crude oil production in US increases from about 5.5 million barrels per day in 2010 to a peak of 6.7 million barrels per day in 2020, then declines to about 6.0 million barrels per day in 2035 in the AEO2012 reference case. Overall, the production level in 2035 is more than 9% higher than the 2010 level. The INFORUM projection shows a steady increase in production, to 5.8 million barrels per day in 2035. Domestic crude oil production decreases to 3.2 million barrels per day in 2035 in the P&G projection.

The growth in consumption of world liquids is almost equally concentrated in OECD and non-OECD countries (Figure 11). Increases in consumption in both regions are quite similar for the different projected years. This can be a result of a established balance between production and consumption from 2015 to 2035 (EIA 2011).

U.S. consumption of petroleum and other liquids totals 19.9 million barrels per day in 2035 in the AEO2012 reference case, an increase of 0.7 million barrels per day over the 2010 total. The transportation sector grows by about 0.6 million barrels per day from 2010 through 2035, while petroleum and other liquids consumption remain relatively flat. The transportation sector accounts for 72% of total petroleum and other liquids consumption in 2035.

3.5. Petroleum and the Environment

Petroleum commonly referred to as crude oil, is a complex mixture of numerous different kinds of hydrocarbons of various molecular weights; many of them are highly toxic and cancer originators (carcinogenic). Because crude oil is toxic to fish, a concentration of 4000 parts per million (ppm) would kill a fish quickly (Prasad). The crude oil and gasoline contains benzene, which has been associated to cause leukemia in humans (Lockslaw, Kirkeleit et al.).

When petroleum distillates are burned in vehicles or stationary sources, generally the combustion is not complete. This means that incompletely burned compounds are produced in addition to water and carbon dioxide. These produced compounds such as carbon monoxide and methanol are often toxic to life. Moreover, fine particulates of soot are deposited in the lungs of humans and other animals leading to heart problems or death. Therefore, soot is classified as carcinogenic.

High temperatures derived from the combustion of petroleum cause the oxidation of nitrogen gas in the surrounding air, creating nitrous oxides. Nitrous oxides, and sulfur dioxide from the sulfur in the oil, combine with water in the atmosphere to generate acid rain. Many problems such as dead trees and acidified lakes with dead fish are consequences from the acid rain. Coral reefs in the oceans of the world are also killed by acid rain.

In the cities, acid rain leads to increased corrosion of machinery and structures, and to the slow destruction of important archaeological structures. Humans burning large amounts of petroleum create large amounts of carbon dioxide (CO_2) that traps heat in the upper atmosphere. Also some organic compounds, such as methane released from petroleum drilling or from the petroleum refining, trap heat several times more than CO_2. Soot blocks the sun from reaching the earth and could cause cooling of the atmosphere, therefore a climate change.

An oil spill, which is another form of pollution, is the release of liquid petroleum into the environment, especially marine areas, due to human activity. The spills are usually observed in the oceans or coastal waters, but they may also occur on land. Oil spills may be due to releases of crude oil from tankers, offshore platforms, drilling rigs and wells, as well as spills of refined petroleum products (such as gasoline, diesel) and their by-products, heavier fuels used by large ships such as bunker fuel, or the spill of any oily refuse or waste oil. Table 5 shows the important spills occurred during 2011 and the averaged spilt volume of oil.

Cleanup and recovery from an oil spill is difficult and depends upon many factors, including the type of oil spilled, the temperature of the water, and the types of shorelines and beaches involved, therefore, spills may take weeks, months or even years to clean up (Holleman, NOAA).

Volatile organic compounds (VOCs) are gases or vapours emitted by various solids and liquids, many of which have short- and long-term adverse effects on human health and the environment. VOCs from petroleum are toxic and foul the air, and some like benzene are extremely toxic, carcinogenic and cause DNA damage. More important for vapors from spills of diesel and crude oil are aliphatic, volatile compounds. Although less toxic than compounds like benzene, their overwhelming abundance can still cause health concerns even when benzene levels in the air are relatively low. The compounds are sometimes collectively measured as "Total Petroleum Hydrocarbons" or "TPH" (HDOH).

Waste oil is used oil containing breakdown products and impurities from use. Some examples of waste oil are used oils such as hydraulic oil, transmission oil, brake fluids, motor oil, crankcase oil, gear box oil and synthetic oil. Many of the same problems associated with natural petroleum exist with waste oil. When waste oil from vehicles drips out engines over streets and roads, the oil travels into the water table bringing with it such toxins as benzene, poisoning both soil and drinking water. Runoff from storms carries waste oil into rivers and oceans, poisoning them as well.

Conclusion

The proved reserves of natural gas distributes around the world, where Eurasia is the highest contributor, represent a good choice to contribute to the balance of the consumption of clean fuels because Natural gas emits to the environment, less contaminants than the other fossil fuels.

It is expected that US will maintain his leadership in Natural gas production after Russia was displaced in 2009. Middle East will maintain its usual equilibrium in production-consumption.

The projections for 2025 and 2035 of natural gas production, consumption, and prices aim to moderately increase, nevertheless the differences in assumptions for the estimations. For this reason, natural gas will continue as the fuel of choice in many regions of the world in the electric power and industrial sectors. Since natural gas is considered the cleanest of the fossil fuels, emission of pollutants to the atmosphere will be reduced in the manner the other fossil fuels are replaced in the burning processes.

Since the crude oil reserves during the last years have slightly increased, this behavior is expected to be unchanged during the next decade. The economy will remain seriously dependent on crude oil, if natural gas or other fuels (fossil or biofuels) are not strongly encouraged to replace that dependence.

Petroleum will remain being the main fuel in the world for many years; however, serious restrictions should be taken to avoid spills and gas leaks caused by human handle. Additionally, constrains on exhaust emissions to the atmosphere should become more stringent the next two or three decades. These constraints should be focused on the inclusion of biofuels in order to improve the environment in an integral behavior.

REFERENCES

AFOP, (2012). Action for our planet. Website: http://www.actionforourplanet.com. Accessed December 2012.

British Petroleum, (2012). BP Statistical review of world energy June 2012. http://www.bp.com/statistical/review. Accessed December 16, 2012.

Cedigaz, 2010 Natural Gas Year in Review: Cedigaz-First Estimates, (April 2011), website www.cedigaz.org/Fichiers/FEstimates2010.pdf. Retrieved December 2012.

Clarke Energy. Coal Gas. www.clarke-energy.com, accessed 16.01.2013.

Devold, H., (2006). Oil and gas production handbook-An introduction to oil and gas production. ABB ATPA Oil and Gas.

EIA, (2012). Annual Energy Outlook 2012 with Projections to 2035, Report #: DOE/EIA-O383, Integrated and International Energy Analysis, Energy Information Administration, Washington, DC.

EIA, (2011). International Energy Outlook, Energy Information Administration, Report #: DOE/EIA-O484, Washington, DC.

EIA, (2006). Annual Energy Outlook 2006 with Projections to 2030, Report #: DOE/EIA-O383, Integrated and International Energy Analysis, Energy Information Administration, Washington, DC.

EIA, (2010). US Energy Information Administration-Independent statistics and analysis, Country analysis briefs. http://www.eia.doe.gov. Retrieved 2010-10-05.

EIA, (1999a). Natural Gas 1998 Issues and Trends, Energy Information Administration, Report #: DOE/EIA-0560, Washington, DC, April.

EIA, (1999b). Petroleum: An energy profile 1999, Energy Information Administration, Report #: DOE/EIA-0545, Washington, DC.

Gallagher, J. E., (2006). Natural gas measurement handbook. Gulf Publishing Company, Houston, TX.

HDOH. "Field Investigation of the Chemistry and Toxicity of TPH in Petroleum Vapors: Implications for Potential Vapor Intrusion Hazards". Hawai'i Department of Health. Retrieved 2012-12-08.

Holleman M., The Lingering Lessons of the Exxon Valdez Oil Spill. The Seattle Times Co. March 22, 2004. Retrieved 2013-01-19.

IEA, (2012). CO2 emissions from fuel combustion-Highlights, International Energy Agency, 2012 Edition, Paris, France.

Kirkeleit J., Riise T., Bratveit M., Moen B. E. "Benzene Exposure on a Crude Oil Production Vessel. *Ann. Occup. Hyg.*, Vol. 50, No. 2, pp. 123–129, 2006.

Lockslaw firm. "Pennsylvania, New Jersey–Philadelphia Toxic Tort/Chemical Injury Attorneys". www.lockslaw.com. Accessed 2013-01-04.

McCoy, Mark; W. Neal Sams (2007). "Tight Gas Reservoir Simulation: Modeling Discrete Irregular Strata-Bound Fracture Networks and Network Flow, Including Dynamic

Recharge from the Matrix". National Energy Technology Laboratory. http://www.netl.doe.gov/kmd/cds/disk28/NGP17.PDF. Retrieved 16 January 2013.

NOAA. Hindsight and Foresight, 20 years after the Exxon Valdez Spill". National oceanic and Atmospheric Administration, NOAA Ocean Media Center. 2010-03-16. Retrieved 2013-01-19.

Prasad M. S. 2006, "Toxicity of Crude Oil to the Survival of the Fresh Water Fish Puntius sophore (HAM). Acta hydrochimica et hydrobiologica-Wiley InterScience". www3.interscience.wiley.com. Consulted 2013-19-01.

Speight, J. G., (2007). Natural gas: a basic handbook. Gulf Publishing Company, Houston, TX.

SPE (2007). Petroleum Resources Management System Guide for Non-Technical Users. http://www.spe.org/industry. Retrieved December 2012.

SPE 2007. "Petroleum Resources Management System". Society of Petroleum Engineers 2007. http://www.spe.org/spe-app/spe/industry/reserves/prms.htm.

SPE (2011). AAPG, WPC, SPEE. http://www.spe.org/spe-app/spe/industry.

SPE (1997). "Petroleum Reserves Definitions". Petroleum Resources Management System. Society of Petroleum Engineers 1997. http://www.spe.org/spesite/spe/spe/industry/reserves/Petroleum_Reserves_Definitions_1997.pdf.

Research and Markets, (2009). Heavy Crude Oil, Outlook to 2020 - Covering Both the Upstream and Downstream Aspects of Heavy Oil." The Free Library. http://www.thefreelibrary.com.

Solari R. B., Marzin R, Zbinden H. Comparison of Carbon Rejection and Hydrogen Addition Processes in Production-Upgrading Complexes. 15th World Petroleum Congress, Beijing, China. October 12-17, 1997.

Vartivarian D, Andrawis H. Delayed coking schemes are most economical for heavy-oil upgrading. *Oil and Gas Journal* 2006;104(6):52-6.

In: Natural Gas and Petroleum
Editor: Constantin Cranganu
ISBN: 978-1-62808-535-8

Chapter 5

AN OVERVIEW ON CRUDE OIL-IN-WATER EMULSIONS AND THEIR APPLICATION IN THE HEAVY OIL PRODUCTION AND TRANSPORTATION

Ronaldo Gonçalves dos Santos*
Laboratory of Miscible Displacement Methods, Center of Petroleum,
University of Campinas, Campinas, SP, Brazil

ABSTRACT

Crude oil-in-water (O/W) emulsions have been proposed as an alternative and effective way to move viscous crude oils from oilfields to refinery and seaports, as well as across porous media inside the reservoir. O/W emulsion technology constitutes a lubricating flow method based on the dispersion of the crude oil as droplets within an aqueous phase containing emulsifying agents for reduction of the frictional energy loss produced during the viscous flow. The formation of a lubricating layer near to the pipe wall leads to a reduction of the frictional pressure drop and consequently to a decrease of the required energy for pumping. O/W emulsion flow represents also a productivity improvement on heavy oil pipeline by means of flow rate increase and the reduction of the maintenance and cleaning costs. This chapter discusses aspects of the formulation and characterization of crude oil-in-water emulsions for moving viscous crude oils. The main goal is to provide an overview of the main properties of petroleum disperse systems related to oil-in-water emulsions. It is also pointed out the effects of fluid dynamic, preparation and compositional variables on the stability, droplet size and viscosity of oil-in-water emulsion and their relevance for oilfield research and application of crude oil in water emulsions. Interfacial and colloidal aspects are highlighted and the role of natural surfactants from the oil on the emulsion properties is balanced. Technological outlook and practical experience in laboratory and oilfields were revised with support of the successfully worldwide research.

Keywords: Heavy oil; O\W emulsion; Pipeline; Oil recovery; Asphaltene; Emulsifier agent

---

* Corresponding Author address: Email:rgsantos@fei.edu.br.

INTRODUCTION

Emulsions are thermodynamically unstable systems attained through the dispersion of two immiscible liquid phases and therefore they have the natural tendency to separate phases and regenerate the original system (Rosen, 1989). This behavior occurs because the dispersion between the phases produces an increase of the free energy of the system, promoting an intrinsic instability (Everett, 1989). For maintaining the system in the disperse form, avoiding the quick phase separation, the addition of one third component is required. The extra component is termed emulsifier. Emulsifiers are surfactants, an acronyms for surface active agent, which presents chemical affinity to both phases and acts on the oil/water interface, reducing the interfacial tension and retarding the phase separation (for details, see Rosen, 1989). In the petroleum industry, surfactants are commonly used as additives in drilling fluids and in enhanced oil recovery methods. The presence of surfactant at the interfaces impacts strongly on foaming, emulsification and wetting properties of petroleum multiphase systems.

Emulsions are sometime termed macroemulsions to contrast with microemulsions, which represent thermodynamically stable disperse systems. Emulsions are classified into two major groups: single emulsions and multiple emulsions. Single emulsions are formed by just one dispersed phase and they are termed either water-in-oil (W/O) or oil-in-water (O/W). W/O emulsions are formed by aqueous phase droplets dispersed in a continuous oil phase. Oil continuous emulsions are easily produced during the simultaneous flow of oil and water under high shear rates provided by the pipeline facilities, mainly chock valve. These emulsions are generally unwanted in oilfield operations because their harmful effects on the flow properties and on the oil quality. O/W emulsions are commonly termed inverse emulsion because they are constituted by aqueous continuous phase systems containing oil phase dispersed as droplets. Multiple emulsions are defined as emulsions in which the disperse phase contains droplets of another immiscible fluid inside, but generally this fluid is the same constituting the continuous phase (Becker, 1997). The emulsion type depends on a wide number of variables, such as composition of water and oil phases, emulsifier properties, temperature and pressure. A generalized qualitative rule – known as Brancroft rule - states that surfactants having higher affinity for aqueous phase tend to form O/W emulsion, while surfactants with higher affinity for oil phase tend to form W/O emulsions (Brancroft, 1913). This affinity can be quantitatively evaluated by the concept of hydrophilic-lipophilic balance (HLB) stated by Griffin (1949) and by its more recent improvements (Binks and Lumsdon, 2000; Salager et al., 2001).

Petroleum emulsions are not simply oil droplets dispersed in water or vice versa, but rather, a sophisticated mixture of chemical structures which the partition between the phases makes the system more stable (Becker, 1997). The ability of crude oil to form direct or inverse emulsions is strongly dependent on the composition of the crude oil (Schramm, 1992) and on the surfactant action of its polar fractions (Buckley, 1994). Oil-in-water emulsions are characterized by their high interfacial density of negative charge and a diffuse layer of attracted surrounding ions, forming a double electric layer (Davies and Rideal, 1963). As a result, these emulsions generally exhibit a complex rheological behavior, attributed to the combined effect of surface charge, the ionic strength of the continuous phase, particle size distribution, volumetric fractions of the phases, dispersed phase viscosity and chemical structure of the emulsifier (Sanchez and Zakin, 1994). The properties of the interfacial film in petroleum emulsions are mainly dependent on the concentration of polar molecules in the oil, the composition and pH of the aqueous phase, the temperature and film compressibility (Kokal et al., 1992).

The use of oil-in-water emulsions in conventional pipelines has been subject of research in many oil companies and research centers because heavy oils are becoming increasingly abundant. They are pointed out as emergent technology for increasing the productivity rate of mature fields and improving the rate of secondary recovery of heavy oil wells. The emulsion technology should be used as an alternative to conventional methods for oil transportation and production since it can reduce the fluid viscosity to an appropriate level for pumping into the current oil pipeline systems. Besides their utilization in the transport process in pipeline and oil reservoirs, O/W emulsions can be employed to burn of heavy fuels.

Concentrated oil-in-water emulsions can be arranged in a lubricated flow standard, in which the disperse droplets migrate to the central tube region, creating a sliding aqueous layer along the walls of the pipe. This lubrication system is similar to the lubricating shear flow of concentrated suspensions of solid particles, considering that the natural migration of particles away from the high shear region (near the wall) creates a concentrated central region. The formation of a lubricating layer near the wall leads to a reduction in frictional pressure drop, which is even greater than the decrease promoted by emulsification. This layer should also sustain the water-wet behavior for the pipeline surface, avoiding oil incrustation and pipe blockage.

STABILITY OF PETROLEUM EMULSIONS

The main characteristic of disperse systems is the large interfacial area between the particles and their surroundings, and consequently a great number of molecules are associated with micro-heterogeneous regions between phases. In these systems, the driving force for aggregation is the free energy gradient in the surface (Everett, 1989). The surface free energy is dependent on the area and surface tension, such that at constant temperature and pressure, surface tension can be derived from the Gibbs free energy, given by equation (1).

$$\gamma = \left(\frac{\partial G}{\partial A} \right)_{T,P} \tag{1}$$

where G represents the Gibbs free energy, A is the surface area and γ is the surface tension. However, the energy required for emulsification is a few orders of magnitude higher than the Gibbs free energy ΔG. The additional energy for the creation of a new interface can be expressed by the Young-Laplace equation (Davies and Rideal, 1963).

$$\Delta P = \gamma \cdot \left(\frac{1}{R_1} - \frac{1}{R_2} \right) \tag{2}$$

Oil/water dispersion tends to minimize the Gibbs free energy by reducing the interfacial area (Kokal et al., 1992), according to the interfacial energy (Davies and Rideal, 1963; Everett, 1989). The minimization of the interfacial area can occur through two mechanisms: (i) coagulation, possibly followed by coalescence, and (ii) Ostwald ripening. For practical purposes, the term "emulsion stability" refers to the kinetic aspects of dispersion resistance in relation to the coalescence of the dispersed droplets. Consequently, the rate of coalescence of the droplets is established as the unique quantitative measurement of the emulsion stability (Boyd et al., 1972). Coalescence can be prevented by the incorporation of suitable emulsifying agents. The driving forces for the mass transference process that leads to Ostwald ripening are the compositional difference in the droplet interface, which is directly related to droplet size distribution. Then, Ostwald ripening, which is controlled by the molecular diffusion of the solute, will only cease with the extinction of the curvature between the phases (Capek, 2004).

Emulsion instability is originated from breaking, flocculation, sedimentation and phase inversion processes (Shaw, 1975; Kokal et al., 1992), which are related to the magnitude of the interfacial forces between the two liquids (Everett, 1989) and to the magnitudes of the kinetic and gravitational effects. The process of breaking emulsions comprises essentially of three main steps: (i) long-distance interaction between the droplets or between the droplets and the surface, (ii) drainage of the interfacial film and finally (iii) coalescence (Tadros and Vincent, 1986; Salager, 2000). Coalescence, which represents the last stage before the phase separation, can be limited by the formation of mechanical, steric or electrical barriers on the interface. In this way, emulsion stability will depend on the [1] physical nature of the interfacial film, [2] existence of electrical charges or steric impairment on the droplet surface, [3] viscosity of the continuous phase, [4] size and distribution of the size of dispersed droplets, [5] volumetric ratio between the phases and [6] temperature (Rosen, 1989).

Long distance interactions are enabled by the Archimedes buoyancy effect, which operates in systems with large density differences between continuous and dispersed phases. These differences cause creaming or settling, both with similar nature to the Stokes sedimentation. Due to the quadratic relationship between the radius of the dispersed particle and the Stokes velocity, dispersion of very small droplets show a low approach velocity between the droplets, and consequently are more stable (Salager, 2000; Capek, 2004).

When the droplets are close to each other, the liquid film that separates them displays a complex drainage process. This drainage process involves mechanisms that depend on droplet volume (involving van der Waals attractive forces and buoyancy), properties of the film (such as interfacial visco-elasticity) and interfacial phenomena that occur at the submicron scale at the interface (Salager, 2000).

Thinning and rupture of the interfacial liquid film that separates the dispersed particles is one of the primary factors for the instability and breaking of emulsions. If collision occurs between the droplets, the liquid film rupture will very likely conduct coalescence. These processes are governed by surface forces that act between the droplets. The thin liquid film that is formed when two droplets become nearer each other is a metastable molecular layer. Its rupture can occur due to mechanical instability, such as that occurring in spinodal decomposition or by nucleation of a thermally activated cavity (Capek, 2004). The rupture of the liquid film is usually due to mechanical or thermal fluctuations caused by surface oscillations generated by surface stretching. In surfactant-containing systems, these oscillations can be dampened by the Marangoni effect (Issacs and Chow, 1992). Also, barriers between two droplets can be achieved by the presence of charges on the surface of the dispersed droplets and the adsorption of macromolecules, causing steric repulsion. On these considerations, the presence of adequate emulsifying agents should stabilize an inherently unstable dispersions by sufficient time for their application, by means of the reduction of o/w interfacial tension and the decrease of the rate of coalescence between the droplets.

INTERFACIAL AND COLLOIDAL ASPECTS OF CRUDE OIL EMULSIONS

The emulsification process is resulting from the dynamic equilibrium between the processes of droplet formation (fragmentation of one phase) and droplet coalescence (Salager, 2000). In both processes the properties of the interfacial film play an important role. Droplet formation occurs through liquid interface deformation and subsequent fragmentation if sufficient energy is provided to compensate the surface energy increase. The energies involved in the equilibrium of the forces come from dynamic pressure, viscous tension or surface tension.

Amphiphilic molecules with pronounced surfactant characteristics are capable of adsorbing on the interface and reducing interfacial tension (Rosen, 1989). A key aspect of these molecules is their ability to stabilize dispersed structures. The adsorption of molecules at the interface results in the formation of molecularly oriented films or layers, whose thickness depends on the orientation of the molecule (Shaw, 1975). Monomolecular films are often ordered in a simple arrangement. Molecular layers are often investigated by Langmuir, Langmuir-Blodgett and Langmuir-Schaefer techniques.

Asphaltenes and resins stand out as the main polar components from crude oil petroleum and comprise the largest proportion of its interfacial active compounds (Nordli et al., 1991; Sheu et al., 1994; Loh et al., 2007). Although polar heterocompounds are concentrated in the asphaltene fraction, the compounds with interfacial activity are not limited to it (Buckley, 1994). Compounds with acid functionality, such as carboxylic acid, are also re-known as having significant interfacial activity (Seifert and Howells, 1969; Rudin and Wasan, 1992; Ovalles et al., 1998) and they can be out of the asphaltene-type molecules (Santos et al., 2006). The presence of naphthenic acids in crude oil affects surface wettability in pipelines (Santos et al., 2006) and the properties of the adsorbed film (Fang et al., 2002). The crude oil molecular species responsible for interfacial activity are not in chemically distinguishable classes, but most have some polar functionality in common (Speight, 1991, Buckley, 1994).

Asphaltenes are commonly defined as the crude oil fraction insoluble in n-alkanes of low molecular weight and soluble in toluene or benzene (Speight, 1991). They represent a class of petroleum macromolecules composed of polycondensated aromatic rings linked to predominantly aliphatic lateral chains, presenting acidic and basic functional groups at a lesser proportion. Asphaltene molecules can self-assembly under changes in the composition, temperature and pressure conditions, producing colloidal-size structures able to precipitate and form flakes with potential to damage the media (Sheu et al., 1994; Mohamed et al, 1999). Resins are less active than asphaltenes and have lower aromaticity. However, they have a great influence on asphaltene solvency in crude oil, controlling the extent of the aggregative process of asphaltene (Speight, 1991; Loh et al., 2007).

The presence of asphaltenes at the liquid-liquid, solid-liquid and liquid-gas interfaces is dependent on the aggregation state (size and shape) of these particles dispersed in the oil (Fordedal et al., 1996; Li et al., 2002). Asphaltene aggregation is related to oil solvency and its interactions with resins and other macromolecules. Acid-base interactions between asphaltenes and other amphiphilic substances are important for the resistance of the interfacial film. The strength of these interactions varies according the acidity amphiphile (Chang and Fogler, 1994) and the characteristics of the asphaltenic fraction (Östlund et al., 2003). Asphaltenes are supposed to interact through acid-base interactions with the naphthenic acids in crude oil (Ostlund et al., 2003).

Asphaltenes and resins tend to interact with interfaces, forming a steric barrier and reducing possibly interfacial tension of some systems, mainly at low concentrations (Kokal et al., 1992). For water-in-oil emulsions, these phenomena lead to stability increasing (Kumar et al., 2001), but generally their stability depends on the asphaltene-resin established interactions and the adsorption of these particles in the water-oil interface. Salou et al. (1998) associated the contents of asphaltene and resins with the stability of bitumen dispersions in acidic water, concluding that high resin/asphaltene ratios cause emulsion stability and low ratios do not preserve stability. However, Schorling et al. (1999) and Kallevik et al. (2000), contrasting Salou et al. (1998), found a reduction in stability upon the increase of the resin/asphaltene ratio.

The surface activity of naphthenic acids in petroleum depends on their ionization form. Ionized acids and asphaltenes reduce the interfacial film stability formed between the oil and an alkaline solution (Fang et al. 2002). However, non-dissociated acids play an important role on the reducing of the oil-water interfacial tension and on the synergistic interactions with other petroleum colloids (Rudin and Wasan, 1992). Naphthenates can be form structures such as liquid crystals with organic solvents if the oil undergoes alkaline treatment (Horváth-Szabó et al., 2003). A film formed by a mixture of ionized carboxylic acids and asphaltenes can be equivalent to a flexible film, whereas a higher rigidity is observed when the film is formed only with asphaltenes (Wu, 2003).

Practical Aspects of Heavy Oil Emulsions

Applications of petroleum emulsions are founded on the basic scientific principles of colloids and interfaces, which govern nature, stability and other emulsion properties (Schramm, 1992).

The technology of oil-in-water emulsions for the transport of heavy oils and bitumen provides a viable alternative to thermal and cold methods. The economic analysis of the emulsion transport system is highly specific and depends on a large variety of factors, but the highest cost is certainly associated with the use of an emulsifier to guarantee stabilization (Rimmer et al., 1992). This cost depends on the chosen formulation, pumping distance, and in some cases, the oil characteristics and emulsion preparation method. In typical formulations, where the emulsifier content is about 500 ppm, the cost of the emulsifier is estimated to be $0.50-1.00 per oil barrel emulsified for pipelines around 300-700 meters long (Rimmer et al., 1992).

The rheological behavior and stability are the main aspects in the characterization of emulsions used for transportation. Emulsions may present either Newtonian or non-Newtonian behavior, depending mainly on the concentration of the dispersed phase (Pal et al., 1992). At low and moderate concentrations, emulsions generally behave as Newtonian fluids. However, at higher concentrations, there is a reduction in viscosity with increasing shear rate (Santos et al., 2011). The dispersion of oil in water can produce systems with viscosities below 200 cP (Brown et al., 1996; Ahmed et al., 1999; Yaghi and Al-Bemani, 2002). The apparent viscosity of an emulsion containing 60% crude oil can be 10,000 times smaller than the original crude oil viscosity (Plegue et al., 1989). The smaller the oil content, the lower the emulsion apparent viscosity. Ahmed et al. (1999) showed a reduction in viscosity in crude oil from 1500 cP to less than 25 cP for emulsions with 50% oil, whereas viscosities greater than 40cP were obtained for a 60% oil content.

Zaki (1997) formulated oil-in-water emulsions for heavy oil transportation using sodium dodecylbenzene sulfonate (SDS) as an anionic commercial surfactant. It was obtained emulsions with up to 90% oil v/v that were kept stable for fourteen days for a 500 ppm of surfactant concentration. These emulaions were prone to be destabilized by the addition of salt (NaCl). Zaki (1997) showed that the stability afforded by an anionic surfactant is due to electrostatic repulsion, which is responsible for reducing the effects of coalescence of the oil droplets.

Yaghi and Al-Bemani (2002) showed that the viscosity of oil-in-water emulsions containing 70% to 75% crude oil can reach feasible values for transport in pipelines at room temperature. The viscosity of these emulsions under high shear (about 1200 s-1) is lower than 130 cP. Similar values are obtained only with the dilution of crude oil with 20-30% light oil and heating at 50 °C. However, emulsification is a much more affordable option. Ahmed et al. (1999) observed that the viscosity of emulsions formulated with non-ionic surfactants (nonylphenol ethoxylate) and containing 70% oil was inversely proportional to the oil content, the velocity of the mixture and salinity of the aqueous phase. The minimum viscosity of the emulsion containing 70% crude oil was obtained at a concentration of 0.25% (mass) of surfactant, although stable emulsions were obtained with only 0.0625% (mass) of surfactant.

Investigating alternative formulations for the elimination of the use of costly commercial surfactants, Sanchez and Zakin (1994) prepared emulsions with 40-60% v/v of ultraviscous oil dispersed in water using sodium hydroxide. The authors suggested that in alkaline conditions, oil-in-water emulsions are stabilized by the action of natural surfactants found in petroleum that contained ionizable carboxylic groups.

OILFIELD PRODUCTION AND TRANSPORTATION

The transport of emulsified heavy oils consists of dispersing the oils in an aqueous solution. The continuous phase determinates the major aspects related to the energy dissipation during the flow. Then O/W emulsions are supposed to provide an improvement in productivity either through an increase on the effective flow rate or by reducing the costs associated with production, maintenance and cleaning, since the oil is transported without necessarily establishing direct contact with the pipeline surface (Rimmer et al., 1992).

The flow of viscous oil in the form of oil-in-water emulsions is an attractive route for the hydraulic transport of heavy oils because emulsification can reduce viscosity to values of 50-200 cP Rimmer et al., 1992), which can be easily pumped (Ahmed et al., 1999; Yaghi e Al-Bemani, 2002; Langevin et al., 2004). Furthermore, emulsion technology can improve residual oil removal from mature oilfields that are not efficiently recovered by traditional methods that apply heat or diluents.

The technological viability of emulsified oil for heavy oil pipeline application was clearly demonstrated by the development of the production pioneer process of fuel emulsions in large-scale, ORIMULSIONS®. ORIMULSIONS® was applied for the generation of energy in thermal plants by the state-owned PDVSA in Venezuela (Salager et al., 2001; Langevin et al., 2004).

Several studies have supported the viability of oil-in-water emulsions as a technology for transportation of viscous oils. Tests performed under flow of a 13°API oil from Shanjiasi (China) show a 80% reduction in pressure loss, which was achieved by emulsions formulated with water fractions around 0.6-0.8 (Zhang et al., 1991). The pressure drop reduction occurred especially when the test were conducted in more drastic conditions, as using extra-viscous oils at low temperatures. A joint project between BP Canada and the Alberta Energy Company has developed TRANSOIL®, a technology that is comprised by oil emulsification, characterization of transport properties, and evaluation of storage and recovery conditions by de-emulsification (Stockwell et al., 1988). The emulsions obtained were pumped continuously for six days at 80 m^3/day and stored for six days without any sign of degradation. The de-emulsification produced 0.5% BS&W (bottom sediment and water) for the heavy oil and less than 100 ppm oily contaminants in the separated water.

Finally, oil-in-water emulsions can be considered for applications in improved recovery processes of heavy oils and for increasing the recovery factor of mature fields, including the amounts not retrieved from the exhausted fields of light oils (Lissant, 1976; Giuggioli and De Ghetto, 1995). Bertero et al. (1994) tested heavy oil emulsification in Sicily and the Adriatic Sea (Italy) for application in transport and production. The system was based on the injection of an emulsifying aqueous phase into the well without any modification to the existing system. Contact with the oil in the well produced oil-in-water emulsions with 70% dispersed oil. The emulsion viscosity was 30-50 times lower than that obtained with dilution and the oil productivity was four times greater if it is compared to conventional production by diesel dilution

Implementation of the technology of oil-in-water emulsions should include the separation of the oil and water phases after transportation and subsequent treatment and disposal. In this case, the technology must take into account the oil-in-water emulsion preparation with properties that allow for crude oil recovery at the end of the pumping process in a simple and economical way.

CRUDE OIL DEMULSIFYING

The use of emulsion technology for the transport of viscous oils in the oil industry assumes that at the end of the process, the phases can be efficiently recovered under economically and environmentally favorable conditions. This fact focuses the attention of petroleum emulsion studies to the process of de-emulsification, in which the kinetic stability needs to be greatly reduced and the phases separated.

Due to its inherent instability, oil-in-water emulsions have the natural tendency to separate the original phases, due to the density difference between the liquids. However, the presence of an emulsifier can guarantee kinetic stability for a very long time. Thus, emulsion breakage involves the destabilization of the interfacial film of emulsifier around the droplets (Grace, 1992). The main methods used to achieve this film destabilization employ treatment in gravitational vessels using a temperature increase, application of specific chemical agents, electrical field application and addition of diluents. The requirements for equipments and for process conditions to de-emulsify the oil are the same or similar to those used in conventional petroleum treatments (Rimmer et al., 1992).

The amount of dispersed water, along with solid particles in suspension and other contaminants present in the aqueous phase affects crucially the quality of crude oil. The removal of water from the oil should take into account the specifications necessary for transport, refining and storage in an attempt to avoid oversized facilities, reduce energy costs and guarantee operational safety, which are all issues related to maintenance (Triggia et al., 2001). Industrially, this process begins in the separator vessels, where the phases are grossly separated by gravity and density differences between the fluids, by agglutination of dispersed particles and by centrifugal forces.

In order to ensure water content within acceptable levels, crude oil must be subjected to further treatment stages. The thermochemical treatment is widely used in offshore fields and it consists in breaking the emulsion by heating (typically 45-60 ° C) in equipment known as wash tanks or handlers. Electrostatic treatment systems are often found in marine systems and they are based on the application of high voltage (15,000 to 50,000 volts). The process is conducted either by electrophoresis, if the electric field is continuous, or by effect of induced dipole, if the field is toggled.

Regardless of the treatment selected, a suitable residence time, a sufficient stirring and the use of de-emulsifier agents are essential for the separation achieves the suitable conditions. The maximum water content accepted in the process of transportation and refining is usually 1% BS W (basic sediments and water, represented by the ratio between the volume of water and sediment and the total volume of the emulsion), as determined by the standard method BS&W - ASTM 96 (Triggia et al. 2001).

CONCLUSION

The major barrier for exploitation of the large reserves of heavy oil available worldwide is related to the high pressure drop that occurs during the flow of viscous oils either in pipe or porous media. The high viscosity can make unattractive the production costs if the pumping energy requirements are not achieved. Traditional technologies to move viscous oils are the most applied in oilfied, despite the scarcity of light oils and the high cost of diluents as kerosene and naphtha. Crude oil in water emulsions represent currently a promise alternative technology to support the viscous oil flow under low pumping energy. Emulsion technology has also been proposed to be applied to increase the recovery factor for both mature and offshore fields.

O/W emulsion properties are strongly dependent on the oil composition and formulation and emulsification process. Then, a complete investigation should be carry out to determine the optimal conditions to guarantee the feasible flowability and the oil recovery, by de-emulsification after flow process. This investigation should include the selection of an emulsifier (normally, a mixture of surfactants), selection of the emulsification method, description of the emulsion stability and viscosity under a wide range of temperature and pressure conditions, and the definition of the de-emulsification method, which should reduce the BSW to conditions encompassed by the commercial requests.

REFERENCES

Ahmed, N. S., Nassar, A. M., Zaki, N. N., Gharieb, H. K. Formation of fluid heavy oil-in-water emulsions for pipeline transportation. *Fuel.* 78 (5) 593-600. 1999.

Bancroft, W. D. The Theory of Emulsification V. J. *Physical Chemistry* 1913, 17, 501-519.

Becker, J. R. Crude oils, waxes, emulsions and asphaltenes. *PennWell.* Oklahoma. 1997.

Beretta, A., Ferrari, P., Galbiati, L., Andreini, P. A. Horizontal oil-water flow in small diameter tubes: Pressure drop. *International Communications in Heat and Mass Transfer,* 24 (2) 231-239, 1997.

Bertero, L., Dilullo, A., Lentini, A., Terzi, L. An innovative way to produce and. transport heavy oil through dispersion in water: laboratory study and field test results. 69th. Annual Technical Conference and Exhibition, New Orleans, LA, September 25-28, 1994.

Binks, B. P., Lumsdon, S. O. Influence of particle wettability on the type and stability of surfactant-free emulsions. *Langmuir* 16, 8622-8631, 2000.

Boyd J, Parkinson, C. S.; Sherman, P. Factors affecting emulsion stability, and HLB concept. *Journal of Colloid and Interface Science*, 41 (2) 359-370, 1972.

Brown, G. E., Hass, G. R., Sell, R. D. Downhole emulsification: viscosity reduction increases production. *The Journal of Canadian Petroleum Technology*, 35 (4) 25-31, 1996.

Buckley, J. S. Chemistry of the crude oil/brine interface. Proceedings of the 3rd International Symposium on Reservoir Wettability and Its Effect on Oil Recovery, Laramie, WY, Sept. 21-23, 1994.

Capek, I. Degradation of kinetically-stable o/w emulsions. *Advances in Colloid and Interface Science*, 107 (2-3) 125–155, 2004.

Chang, C. L., Fogler, H. S. Stabilization of asphaltenes in aliphatic solvents using alkylbenzene-derived amphiphiles. 1. Effect of the chemical structure of amphiphiles on asphaltene stabilization. *Langmuir* 10 (6) 1749-1757, 1994.

Davies, J. T.: Rideal, E. K. *Interfacial phenomena.* 2.Ed New York: Academic Press, 1963.

Everett, D. H. Basic Principles of Colloid Science. *The Royal Society of Chemistry.* 1989.

Fang, H. B, Zhang, L., Luo, L., Zhao, S., An, J., Xu, Z., Yu, J., Ottova, A., Tien, H.T. A study of thin liquid films as related to the stability of crude oil emulsions. *Journal of Colloid and Interface Science*, 238 (1) 177–182, 2002.

Fordedal, H., Midttun, O., Sjoblom, J., Kvalheim, O. M., Schildberg, Y., Volle, J. A multivariate screening analysis of w/o emulsions in high external electric fields as studied by means of dielectric time domain spectroscopy II. Model emulsions stabilized by interfacially active fractions from crude oils. *Journal of Colloid and Interface Science*, 182 (1) 117–125, 1996.

Giuggioli, A., De Ghetto, G. Innovative technologies improve the profitability of offshore heavy oil marginal fields. International Meeting on Petroleum Engineering: Beijing, 14-17 November 1995.

Grace, R. Commercial emulsion breaking. *Advances in chemistry series* 231, 313-339, 1992.

Horváth-Szabó, G., Masliyah, J. H., Czarnecki, J. Emulsion stability based on phase behavior in sodium naphthenates containing systems: Gels with a high organic solvent content. *Journal of Colloid and Interface Science*, 257 (2) 299–309, 2003.

Isaacs, E. E.; Chow, R. S. Practical aspects of emulsion stability. In Advances in Chemistry Series 231, L. L. Schramm, ed. *American Chemical Society*, Washington DC, 1992.

Kallevik, H., Kvalheim, O. M., Sjöblom, J. Quantitative determination of asphaltenes and resins in solution by means of near-infrared spectroscopy. Correlations to emulsions stability. *Journal Colloid and Interfaces Science*, 225 (2) 494-504, 2000.

Kokal, S. L.; Maini, B.; Woo, R. Flow of emulsions in porous-media. In *Advances in Chemistry Series* 231, 219-262, 1992.

Kumar, A., Nikolov, A. D., Wasan, D. T. Mechanism of stabilization of water-in-crude emulsions. *Industrial & Engineering Chemistry Research*, 40 (14) 3009-3014, 2001.

Langevin, D., Poteau, S., Hénaut, I., Argillier, J. F. Crude oil emulsion properties and their application to heavy oil transportation. *Oil & Gas Sci. and Technol.*, 59, 511, 2004.

Li, M., Xu, M., Ma, Y., Wu, Z., Christy, A. A. Interfacial film properties of asphaltenes and resins. *Fuel*, 81 (14) 1847-1853, 2002.

Lissant, K. J. Emulsification and de-emulsification in oil recovery. In: Improved oil recovery by surfactant and polymer flooding. Shah, D. O. and Schechter, R. S., Eds. Academic Press, UK, 1997.

Liu, Y., Buckley, J. S. Evolution of wetting alteration by adsorption from crude oil. *SPEFE*, 5-11, March 1997.

Loh, W.; Mohamed, R. S.; Santos, R. G. Crude oil asphaltenes: Colloidal aspects. In: P. Somasundaram. (Org.). *Encyclopedia of Colloid and Surface Science*. Nova York: Taylor & Francis, 2007.

Mohamed, R. S., Ramos, A. C. S., Loh, W. Aggregation behavior of two asphaltenic fractions in aromatic Solvents. *Energy & Fuels* 13 (2) 323-327, 1999.

Nordli, K. G., Sjoblom, J., Kizling, J., Stenius, P. Water-in-crude oil-emulsions from the Norwegian continental-shelf .4. Monolayer properties of the interfacially active crude-oil fraction. *Colloids and Surfaces*, 57 (1-2) 83-98, 1991.

Ostlund, J. A., Nyden, M., Auflem, I. H., Sjoblom, J. Interactions between asphaltenes and naphthenic Acids. *Energy & Fuels*, 17 (1) 113-119, 2003.

Ovalles, C., Garcia, M. D., Lujano, E., Aular, W., Bermudez, R., Cotte, E. Structure-interfacial activity relationships and thermal stability studies of Cerro Negro crude oil and its acid, basic and neutral fractions. *Fuel*, 77 (3) 121-126, 1998.

Pal, R., Yan, Y., Masliyah, J. Rheology of emulsions. In *Advances in Chemistry Series*, 231,131-170 1992.

Plegue, T. H., Frank, S. G., Fruman, D. H., Zakin, J. L. Concentrated viscous crude oil-in-water emulsions for pipeline transport. *Chemical Engineering Communications*, 82, 111-122, 1989.

Rimmer, D. P., Gregoli, A. A., Hamshar, J. A., Yildirim, E. Pipeline emulsion transport for heavy oils. In Emulsions: Fundamentals and Applications in the Petroleum Industry. L.L. Schramm, Ed. *American Chemical Society*, Washington, DC, 1992.

Rosen, M. Surfactants and interfacial phenomena. *John Wiley & Sons*. New York. 1989.

Rudin, J., Wasan, D. T. Mechanisms for lowering of interfacial tension in alkali/acidic oil systems: Effect of added surfactant. *Industrial & Engineering Chemistry Research*, 31 (8) 1899-1906, 1992.

Salager J. L. Emulsion properties and related know-how to attain them, Cap. 3 de Pharmaceutical emulsions and suspensions, *F. Nielloud & G. Marti-Mestres*, (Eds), M. Dekker, New York, 2000.

Salager, J. L., Briceno, M. I., Brancho, C. L. Heavy. Hydrocarbon Emulsions, In *Encyclopedic Handbook of Emulsion Technology,* Sjöblom, J. (ed), Marcel Dekker, New York, 2001

Salou, M., Siffert, B., Jada, A. Study of the stability of bitumen emulsions by application of DLVO theory. *Colloids and Surfaces A*, 142 (1) 9-16, 1998.

Sanchez, L. E., Zakin, J. L. Transport of viscous crude as concentrated oil-in-water emulsions. *Industrial & Engineering Chemistry Research*, 33 (12) 3256-3261, 1994.

Santos, R. G.; Bannwart, A. C.; Briceno, M. I.; Loh, W. Physico-Chemical Properties of Heavy Crude Oil-in-Water Emulsions Stabilized by Mixtures of Ionic and Non-Ionic Ethoxylated Nonylphenol Surfactants and Medium Chain Alcohols. *Chemical Engineering Research & Design*, 89 (7) 957-967, 2011.

Santos, R. G., Mohamed, R. S., Bannwart, A. C., Loh, W. Contact angle measurements and wetting behavior of inner surfaces of pipelines exposed to heavy crude oil and water. *Journal of Petroleum Science and Engineering* 51 (1-2) 2006.

Schorling, P. C., Kessel, D. G., Rahimian, I. Influence of the crude oil resin/asphaltene ratio on the stability of oil/water emulsions. *Colloids and surfaces* A, 152 (1-2): 95-102, 1999.

Schramm, L. L. Petroleum emulsions - Basic principles. In Emulsions: Fundamentals and Applications in the Petroleum Industry. L. L. Schramm, Ed. *American Chemical Society*, Washington, DC, 1992.

Seifert, W. K and Howells, W. G., Interfacially active acids in a California crude oil. Isolation of carboxylic acids and phenols. *Analytical Chemistry*, 41 (4) 554-562, 1969.

Shaw, D. J. *Introdução à Química dos Colóides e de Superfícies.* Edgard Blucher. São Paulo, Brasil. 1975.

Sheu, E. Y., De Tar, M. M. and Storm, D. A. Surface Activity and Dynamics of Asphaltenes. In: *Asphaltenes Particles in Fossil Fuel Exploration, Recovery, Refining, and Production Processes.* Sharma, M. K., Yen, T. F., Eds.; Plenium Press: New York, 1994; 115-122. Speight, J. G. The Chemistry and Technology of Petroleum. Marcel Dekker, Inc. 2nd ed., 1991.

Stockwell, A., Sit, S. P., Hardy, W. A. Transoil technology for heavy oil transportation. Results of field trials at Wolf Lake. *Paper SPE* 18362 presented at SPE European Petroleum Conference, London, UK, October 16–19, p. 249, 1988.

Tadros, T. F., Vincent, B. Emulsion stability. In *Encyclopedia of Emulsion Technology,* Becher, P. ed. Marcel Dekker, New York, Volume 1, Chapter 3, pp. 1-56, 1983.

Triggia, A. A., Correia, C. A., Filho, C. V., Xavier, J. A. D., Machado, J. C. V., Thomas, J. E., Filho, J. E. S., Paula, J. L., Rossi, N. C. M., Pitombo, N. E. S., Gouveia, P. C. V. M.,

Carvalho, R. S., Barragan, R. V. *Fundamentos de Engenharia de Petróleo,* Rio de Janeiro: Interciência. 2001.

Wu, X. Investigating the stability mechanism of water-in-diluted bitumen emulsions through isolation and characterization of the stabilizing materials at the interface. *Energy & Fuels,* 17 (1) 179-190, 2003.

Yaghi, B. M., Al-Bemani, A. Heavy crude oil viscosity reduction for pipeline transportation. *Energy Sources*, 24 (2) 93-102, 2002.

Zaki, N. Surfactant stabilized crude oil-in-water emulsion for pipeline transportation of viscous crude oils. *Colloids and Surfaces* A, 125 (1) 19-25, 1997.

Zhang, J., Chen, D., Yan, D., Yang, X., Shen, C. *Pipelining of heavy crude oil as oil-in-water emulsions. SPE* 21733,. 911-918, 1991.

In: Natural Gas and Petroleum
Editor: Constantin Cranganu

ISBN: 978-1-62808-535-8

Chapter 6

BTEX: Toxic Effects and Possible Decontamination by Microbial Biodegradation

Dânia Elisa Christofoletti Mazzeo [*]
and Maria Aparecida Marin-Morales
UNESP—Univ Estadual Paulista, Department of Biology,
Institute of Biosciences, Rio Claro, SP, Brazil

Abstract

Petroleum and its derivatives have been considered important environmental contaminants. Among the petroleum derivatives, the monoaromatic hydrocarbons benzene, toluene, ethylbenzene and xylene (BTEX) are widely used as solvents and raw material in several industrial processes, moreover they are present in large amounts in fossil fuels. Due to their broad use and high solubility in water, BTEX is characterized as a frequent pollutant of waters, especially groundwater. Knowledge about the toxic effects associated to the exposure to the BTEX mixture is quite limited and this information refers mainly to the isolated effects of each one of these substances. In animals, these chemicals are easily absorbed, extensively metabolized and distributed to tissues highly vascularised and rich in lipids. Prolonged exposure to BTEX leads to renal, hepatic and lung toxicity, as well as alterations in the nervous and circulatory system. In certain concentrations, BTEX is genotoxic and mutagenic. The application of the microbial biodegradation process seems to constitute a promising methodology for the treatment of waters contaminated by this mixture. The use of bioindicators for the monitoring of environmental contamination has been considered quite satisfactory for identifying and assessing harmful effects of xenobiotics on test organisms, in their different concentrations and exposure periods, since these data can be extrapolated for organisms of the natural biota of impacted environments. Bioindicators also assist in the evaluation of the biodegradation process of environmental chemicals, allowing observing the effects caused by possible metabolites formed during this process. In this review, it will be presented studies related with the general characteristics of benzene, toluene,

[*] Corresponding Author address. E-mail: daniamazzeo@gmail.com.

ethylbenzene, toluene and xylene and the biological action of these substances on exposed organisms. It will be also discussed the possibility of microbial biodegradation of BTEX and its application for remediation of groundwater contaminated by these xenobiotics.

Keywords: Petroleum derivatives, Monoaromatic hydrocarbons, Volatile organic compound, Bioremediation, Microorganisms, Bioassays

INTRODUCTION

Petroleum, as well as its derivatives, has been considered one of the main contaminants of the environment. Due to great increase in the scale of production, transport and use worldwide, both as source of energy and production of several products such as plastics, solvents, pharmaceutical products, cosmetics, fossil fuels, synthetic rubbers and several others, many researchers have devoted special attention to the investigation of the possible impacts that these products can promote in the environment.

Environmental contaminations by petroleum and its products are commonly caused by accidents in the process of loading, unloading, transportation, production of by-products and combustion (Pedrozo et al., 2002). When these products reach the soil, they can go three distinct ways: remain trapped in porous spaces of the soil; dissolve in groundwater; evaporate and reach the atmosphere (Nadim et al., 2000), presenting therefore a potential for contaminating several physical media of the environment.

Petroleum derivatives such as benzene, toluene, ethylbenzene and xylene are compounds that have aroused a special interest among researchers, due to their toxic characteristics; mobility in the environment, both in the gaseous, liquid and solid phase; and high solubility of the mixture in water (Lovley, 1997; Anneser et al., 2008).

These chemicals have been used in the chemical industry to prepare solid catalysts, in polymerization reactions, to dissolve photosensitive compositions and for synthesizing pesticides and pharmaceutical products (Barreto et al., 2007). They represent about 35% of the composition of fossil fuels, and are present particularly in automotive gasoline, airplane fuels and kerosene and also in a lesser extent in diesel. Moreover, they are produced on the scale of megatonnes per year, as chemical products for industrial use, being used as solvents and raw materials in several products, such as paints and coatings (ATSDR, 2004) and in the manufacture of plastics and synthetic fibres (Hutchins et al., 1991).

Due to the high concentration of the mixture BTEX in petroleum and to the massive use of this mixture as source of energy, as solvents and in the production of other organic chemicals, the contamination by this compound has increased, which can result in serious problems to the environment and public health (Walden; Spence, 1997).

General Characterization of BTEX

The acronym BTEX is formed from the names of the chemical substances: benzene, toluene, ethylbenzene and xylenes (o-xylene, m-xylene, p-xylene). As these substances are generally found together in the environment, the expression BTEX is used as a joint reference to them.

Table 1. Chemical structures and physic-chemical properties of BTEX

	BENZENE	TOLUENE	ETHYLBENZENE	ORTHO-XYLENE	META-XYLENE	PARA-XYLENE
SYNONYMS	Benzol Cyclohexatriene	Methylbenzene Toluol Phenylmethane	Ethylbenzol Phenylethane	o-Xylol 1,2- Dimethylbenzene	m-Xylol 1,3- Dimethylbenzene	p-Xylol 1,4- Dimethylbenzene
CAS NUMBER	71-43-2	108-88-3	100-41-4	95-47-6	108-38-3	106-42-3
ABBREVIATION	B	T	E	X	X	X
MOLECULAR FORMULA	C_6H_6	C_7H_8	C_8H_{10}	C_8H_{10}	C_8H_{10}	C_8H_{10}
STRUCTURE FORMULA		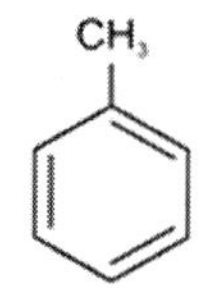	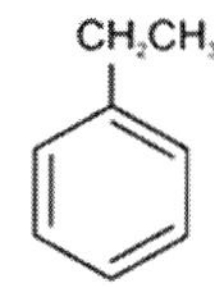	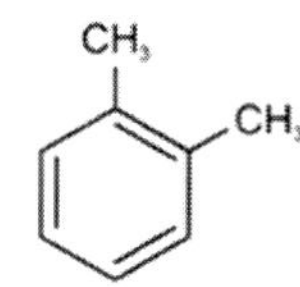	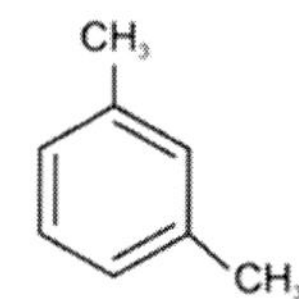	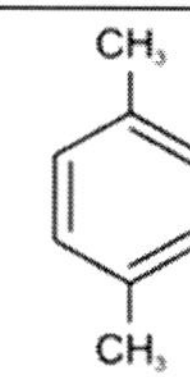
MOLECULAR WEIGHT	78.11	92.13	106.17	106.17	106.17	106.17
DENSITY (g/L a 20°C)	0.8794	0.8660	0.8665	0.8800	0.8640	0.8611
WATER SOLUBILITY (mg/L - 25°C)	1790	526	169	178	161	162
COEFICIENTE DE PARTIÇÃO OCTANOL-ÁGUA (log K_{ow})	2.13	2.73	3.15	3.12	3.20	3.15
MELTING POINT (°C)	5.5	-95.0	-95.0	-25.0	-47.9	-14.0
BOILING POINT (°C)	80.1	110.4	136.25	144.4	139.0	138.3
VAPOR PRESSURE (mmHg - 25°C)	94.8	28.4	9.6	6.61	8.29	8.84
K_{OC} (mL/g)	62	140	200	240	196	311

BTEX belongs to a chemical group called Volatile Organic Compounds (VOCs). Since they present a characteristic of high volatility, they contaminate the atmosphere instead of persisting on the surfaces of waters and soils. These compounds are pollutants that contribute to the processes of environmental degradation, interfering, for example, in the reduction of the stratospheric ozone and in the formation of tropospheric ozone. The oxidation of VOCs in the presence of nitrogen oxides leads to the formation of photochemical smog, harmful to human beings, animals and vegetation (Falcó; Moya, 2007). Regarding water, VOCs are among the pollutants most commonly found in groundwater (Safarova et al., 2004). These hydrocarbons have high mobility in soil-water systems, due to their smaller partition coefficient between octanol-water.

Octanol (2-Ethyl Hexanol - $C_4H_{10}O$)-water partition coefficient (K_{OW}) is a measure of the lipophilicity of a compound, which is defined as the ratio of the concentration of the compound, when it is at equilibrium, after dissolution in a two phase solution, formed by two immiscible solvents, water and octanol. It represents the balance between the hydrophilic and lipophilic properties of the compound and influences the transport of an organic compound in the environment (Silva; Ferreira, 2003). A smaller partition coefficient implies in a slow absorption into the soil and, consequently, a preferential transport via water, favouring the contamination of reservoirs by migrating faster by water (Nakhla, 2003). These compounds, frequently, occur at trace levels, as a result of their volatility, but are found at a much higher concentration in groundwater and are considered as priority contaminants to be evaluated in water analyses (Falcó; Moya, 2007).

The mixture BTEX represents a homogeneous group of volatile organic hydrocarbons with similar physico-chemical properties (Table 1). Benzene, toluene, ethylbenzene and xylenes (ortho, meta and para-xylene) are monoaromatic hydrocarbons, colourless and with characteristic odour, are soluble in water but generally miscible with other organic solvents such as ethyl ether, acetone, ethanol and chloroform (Dean, 1978). The environments contaminated with these chemicals include air, water and soil (ATSDR, 2004).

In the atmosphere, BTEX contamination is related to the evaporation of these compounds present in paints, adhesives, degreasing agents, aerosols (Esteve-Turrillas et al., 2009; Zhang et al., 2012) and, mainly, by burning fossil fuels during urban traffic (Wang et al., 2002; Hinwood et al., 2007; Miller et al., 2012). However, in open environments, this mixture is easily dissipated. In the soil, significative levels of BTEX are mainly found near gas stations, resulting from leakage of gasoline that is stored in underground tanks (An, 2004). Due to their volatility, BTEX is rarely found in surface waters, but they are present, frequently, in groundwater, as a result of leaks of storage tanks and subterranean pipelines (Johnson et al., 2003), besides industrial activities. The limits for the concentration of these compounds in groundwater are quite restricted, due to their high toxicity. The total maximum concentration allowed for drinking water for benzene, toluene, ethylbenzene and xylenes is 10μg/L, 700μg/L, 300μg/L and 500μg/L, respectively (WHO, 2011).

Among the petroleum hydrocarbons, benzene is one of the most soluble compounds in water. In the environment, it is found in the atmosphere mainly due to its presence in exhaust gases of gasoline engines, to the very process of its production, as well as due to the industrial use as intermediate product. In water, it appears in discharges of industrial effluents or by leakages (Irwin et al., 1997). This chemical has been widely used as solvent in pharmaceutical industries, in paints, glues and derivatives; however, recently, there is a strong pressure to restrict this wide use, due to the proven evidences of its carcinogenic characteristics (Roma–Torres et al., 2006).

Toluene is a volatile organic compound of the monoaromatic hydrocarbons class (Mackay et al., 1992), considered as a potential pollutant by USEPA (1991; 1995 a; b). As ethylbenzene and xylenes, toluene is an alkylbenzene. Toluene differs from benzene by containing a methyl group added to the benzene ring (Irwin et al., 1997). This compound is widely used as solvent of paints, adhesives and in pharmaceutical products, and is also present in the gasoline, together with benzene and xylenes (Roma-Torres et al., 2006).

Ethylbenzene is characterized by being an aromatic hydrocarbons present in petroleum and is one of the components of liquid fuels. It is used as solvent in the production of synthetic rubber and styrene. This compound is commonly found in the air in the form of vapour, easily moving from air to water and from air to soil. Once in the atmosphere, the molecule of ethylbenzene reacts, by photolysis, with other chemical products, forming substances that constitute smog. On the surface of water, ethylbenzene reacts naturally with other compounds present in water, while in the soil, great part of ethylbenzene is degraded by bacteria. As ethylbenzene binds moderately to soil, it can also move through the soil particles, contaminating groundwater (ATSDR, 1999).

Xylene is an aromatic hydrocarbon, composed by two methyl groups linked to the benzene ring, presenting three different isomeric forms: ortho, meta and para, and the mixture of these forms is used in a variety of solvents, mainly in the paint industry (Roma-Torres et al., 2006). Xylene tends to evaporate quickly when exposed to the atmosphere, but can be more persistent in sediments, in soil or in groundwater, where it is not found directly exposed to air. Xylene tends to migrate to groundwater and persist in it for months or even years (ATSDR, 1993).

Effects of BTEX on Organisms

Humans can be exposed to BTEX by ingestion or by inhalation. Acute exposure to this class of compounds is associated to skin irritation, depression of the central nervous system and effects on the respiratory system. Prolonged exposure to BTEX can also affect organs, such as kidneys and liver, and the circulatory system (Mazzuco, 2004), and is also considered dangerous for its potential to cause leukaemia (Corseuil; Alvarez, 1996). According to Irwin et al. (1997), the chronic effects of BTEX include, in humans, damages in the liver, kidneys, heart, lungs and nervous system.

Toxicokinetic studies in humans and animals show that these chemicals are well absorbed, extensively metabolized and distributed to tissues rich in lipids and highly vascularised, such as the brain, bone marrow and adipose tissue, due to the high lipophilicity that they present. However, they do not persist in the body for long periods because they are easily eliminated (ATSDR, 2004). Studies with animals and human show a rapid absorption of these compounds by pulmonary via, with retention rates for man between 30 to 80%. In relation to the oral via, absorption of BTEX is fast and efficient, between 80 to 97% of the ingested dose. Absorption of these hydrocarbons vapours by skin is less significant than by pulmonary and oral via, although skin absorption, given their liquid form, is dependent on the hydration and thickness of the corneal layer and blood perfusion rate of the dermis (Pedrozo et al., 2002).

Regarding the biotransformation process, BTEX is biotransformated by the oxidase via of the cytochrome P-450, suffering modifications both in the aromatic carbon and in the alkyl groups of the side chain. These oxidized intermediates are conjugated with endogenous substrates, such as glicuronic acid, sulphates, gluthathione, cysteine or glycine, making them able to be excreted by kidneys or, in some cases, by bile. Biotransformation of these compounds can result in less toxic compounds (hippuric acid) than the actual chemical (toluene) or in products with an even more toxic effect (hydroquinone) than the original (benzene) (Pedrozo et al., 2002).

Workers occupationally exposed to petroleum derivatives with carbon equivalent between 5 and 9, in which BTEX are included, can present alteration of the liver enzymes, expressed by the increase in the serum levels of transaminases, besides an increase in the liver, fatty infiltration and liver insufficiency, resulted from the chronic hepatotoxic action of these substances (ATSDR, 1995; Góes, 1997).

Roma-Torres (2006) observed an increase in the frequency of chromosome aberrations and micronuclei in lymphocytes of workers occupationally exposed to BTX (benzene, toluene and xylene). According to Chen et al. (2008), BTEX can be also considered genotoxic for human lymphocytes maintained in culture, since induce a variety of damages in the DNA, such as single strand breaks, double strand breaks and modification of the oxidative base, detected by the comet assay. The induction of damages in the DNA was also observed by Mazzeo et al. (2013) in cultured cells of rat hepatoma exposed to different concentrations of the BTEX mixture. The micronucleus test showed a mutagenic effect for the concentrations near the solubility of their compounds in water and the comet assay showed a significant genotoxic effect, even for concentrations below the solubility limit.

As for the effects of BTEX in plants, this mixture proved to be toxic for the species *Sorghum bicolor* and *Cucumis sativus* in concentration above 90 mg BTEX/kg of dry soil, inhibiting the germination of their seeds (An, 2004). Genotoxic (chromosome aberrations) and mutagenic effects (micronucleus) were observed for meristematic cells of *Allium cepa,* when these cells were subjected to germination in aqueous solutions of BTEX, whose concentrations were lower than the solubility of their components in water (Mazzeo et al., 2011).

Knowledge about the mechanism of action and toxic effect of the combined action of benzene, toluene, ethylbenzene and xylenes, as well as the additive, antagonistic or synergistic effect of these substances on target organisms is quite limited and the information on the interaction of these chemicals is restricted, mainly, to binary combinations of them (ATSDR, 2004; An, 2004).

Biological Effects of Benzene

Benzene is a potentially carcinogenic pollutant (USEPA, 1986; 1991). Chronic exposure to benzene can result in serious haematological effects for man, in which all blood components can be affected (WHO, 1993; 2000). Exposure to high concentrations of benzene leads to progressive degradation of bone marrow, aplastic anaemia, leukaemia, as well as immunological alterations (Yardley-Jones et al., 1990). In a review on the topic performed by

Zhang et a. (2010) the haematotoxic effect occurs due to the destruction of the progenitor myeloid cells and that this can be considered a phenotypic effect of the exposure to benzene, even at concentrations below 1 mg/L.

Chromosomal damages have been found among workers exposed to very low levels of benzene (Green, 1988), and these results are confirmed in tests with other animals (USEPA, 1985). Occupational exposure to benzene has been associated with a high frequency of chromosome aberrations in peripheral lymphocytes (Dabney, 1994). Some authors affirm that the increase in the chromosome aberrations rate is dose and time dependent of the exposure to benzene (Tompa et al., 1994; Smith; Rothman, 2000).

Countless studies of *in vitro* genotoxicity have shown that benzene is weakly mutagenic or not mutagenic for mutation assays with standardized genes, but is able to induce chromosome breaks and can interfere in the chromosome segregation (Dean, 1985; Waters et al., 1988; Bird et al., 2005). According to Dean (1985), benzene is capable of inducing both structural and numerical (aneuploidies) chromosome aberrations in cells maintained in culture.

Studies performed by Whysner (2000) indicated that benzene induced micronuclei in bone marrow cells of mice, when tested *in vivo*, not confirmed in *in vitro* tests. These data, according to the author, showed that for benzene act is necessary a metabolic activation. Now, when the *in vitro* tests were carried out with metabolites of benzene, they proved to be positive. According to Whysner et al. (2004), benzene induced significative chromosome alterations in lymphocyte culture, but not in CHO cells (Chinese hamster ovary), possibly by the lack of ability to metabolize of the last ones.

In vivo studies conducted on the mechanism of action of benzene in cells of the bone marrow and spleen of mice showed that the formation of micronucleus is resulted both from chromosome break and loss. Thus, the authors concluded that exposure to benzene causes both clastogenic and aneugenic effects and that the type of action can vary depending on the target organ or the cell type (Chen et al., 1994). However, several studies showed the prevalence of the clastogenic capacity of benzene, although unstable DNA adducts, oxidative damages and disturbances in the spindle are also responsible for the toxic effect of this chemical (Whysner et al., 2004).

Studies carried out by Angelini et al. (2011) showed an increase in the frequency of micronuclei in lymphocytes of traffic policemen occupationally exposed to a daily rate of 10 to 15 $\mu g/m^3$ of benzene, showing the potential of this compound in inducing mutagenic damages, besides also being characterized as a product with high environmental risk.

According to Bird et al. (2005), benzene produces cytogenetic effects in animals and humans, including damages to the chromosomes and/or to the division process. Au et al. (1991) observed a significant increase in the frequency of chromosome aberrations in lymphocytes of CD-1 mice exposed to low levels of benzene (40 and 100 μg/L).

Mullin et al. (1998) investigated the frequency and the pattern of mutations in the lacI transgene in lung tissues of Big Blue mice exposed to benzene. These authors concluded that benzene is a mutagenic substance for lung cells, since, when these cells were exposed to vapours of benzene, they presented an increase in the frequency of mutated genes and an increase in the extent of deletions, in relation to animals not exposed to this chemical.

The use of cytogenetic markers, observed by the methodology of *in situ* hybridization (FISH), showed that benzene induced aneuploidy in human metaphase cells (Bird et al., 2005). The authors observed that the chromosomes 5 and 7 of the human karyotype seem to be more sensitive to losses resulted from the action of this product than the other chromosomes, although, according to the authors, this datum was not properly proven in these assays.

Assays with staminal filament cells of *Tradescantia*, performed by Van't Hof and Schairer (1982), showed that the concentration of 4000 mg/L of benzene, in its gaseous form, promoted a significative increase in the frequency of mutations in these plants, showing the mutagenic potential for this test organism. Benzene also induced a significative increase in the frequency of chromosome aberrations of root cells of *A. cepa*, exposed to the concentration of 100 μM (Rank; Nielsen, 1994).

Nowadays, the main interest of studies on benzene is on the assessment of the continuous and prolonged exposure to low levels of this compound, both occupationally and in the environment (Duarte-Davidson et al., 2001; Sul et al., 2002). Studies of the gene expression have identified 4 genes as potential biomarkers of exposure to high doses of benzene, which are related with apoptosis and lipid metabolism. However, in studies that evaluated the changes in the expression of genes of organisms exposed to low concentrations of this chemical (<1mg/L), it was observed that more than 70 genes presented a significant altered gene expression, including genes related to immune and anti-inflammatory responses (Zhang et al., 2010), showing that in these concentrations there is still the need for further studies to understand the real effect of benzene on biological systems.

Mechanisms of the toxic action of benzene are still not fully clarified, although it is believed that they are related to process of biotransformation and a resulting formation of metabolites such as phenol, catechol, muconaldehyde and hydroquinone (Pedrozo et al., 2002). According to the Pan American Health Organization (OPS, 1996), toxicity of benzene is related to the binding capacity of its metabolites with DNA or RNA, which interferes in the repair process and cell replication, resulting in chromosome damages and carcinogenesis. Initial studies of epigenomic showed that low levels can induce hyper-methylation of the p-15 gene and hypo-methylation of the LINE-1 element (Long Interspersed Nuclear Element-1), however, other studies are required to understand the mechanisms by which these type of alterations occur (Bollati et al., 2007).

Biological Effects of Toluene

Toluene is an inhibitor of the activities of the central nervous system (NLM, 1994) and promoter of other effects in this system, such as memory loss and progressive damages to the brain and nerves, both in acute and chronic exposures (Dabney,1994). Acute intoxication by toluene can be expressed by the toxic encephalopathy syndrome, whose characteristics are given by symptoms such as euphoria, emotional and motor instability and language disorders in an initial stage, followed by drowsiness and impairment of the motor and intellectual functions (Pedrozo et al., 2002). Exposures to high concentrations of toluene (between 5000 to 30000 mg/L) can lead, in few minutes, to a neurological condition of disorientation, hallucinations, loss of motor coordination, tremors, convulsions and unconsciousness, which can reach up to coma and death (OPS, 1996; Aitio; Wada, 1993).

Toluene is occasionally detected in drinking water samples at levels generally below 3 μg/L (Irwin et al., 1997). Toluene is totally absorbed by the gastrointestinal tract and is rapidly absorbed, preferably in adipose tissue. It is rapidly metabolized and, after conjugation, is preferably excreted via urine (WHO, 2011).

Studies carried out by the University of Washington (1990) described a minimal or small teratogenicity for toluene. However, according to WHO (2000; 2011), toluene can induce embryotoxicity and foetotoxicity, which can induce delay in the development, abnormalities in the skeleton, low birth weight and neurotoxic damages.

Roberts et al. (2007) subjected pregnant rats, with 6-15 days of gestation, to the concentrations of 0, 250, 750, 1500 and 3000 mg/L of toluene for 6 h/day. It was observed a dose-dependent response by the authors for alterations in the maternal activity such as ataxia, hyper-responsivity, increased water uptake and inhibited food intake increasing, establishing a NOAEL (No-Observed-Adverse-Effect Level) of 750 mg/L and a LOAEL (Lowest-Observed- Adverse-Effect Level) of 1500 mg/L for material toxicity. Regarding the toxicity of the foetus, it was observed a reduction in foetal weight and alterations in the ossification without, however, being found a dose response effect.

For the mutagenicity of the toluene, EC (1993) cites that although there are studies with microorganisms and mammals, the results did not show a mutagenic effect for the compound. But this agency (EC) states that the assays performed were not conclusive for this assertion. It was also registered controversial results for assays of chromosome aberrations performed with toluene for these same organisms (Dabney, 1994).

Studies carried out by Hammer (2002) showed a significative increase in the sister chromatid exchanges in blood lymphocytes of workers occupationally exposed to concentrations of 141 to 328 mg/m^3 of toluene, confirming the genotoxic effect of this substance.

According to Murata et al. (1999), oxidative damages in the DNA, promoted by toluene and its metabolites, identified by the use of probes (^{32}P-5-end-labeled) of DNA fragments of human genes, are important factors in the processes of induction of carcinogenicity and in the reproductive toxicity, although the carcinogenic potential of toluene is lower than of the benzene.

As reported by USEPA (2005) in a toxicological review on toluene, this compound did not present positive effect for the reverse mutation tests with the bacterium *Salmonella typhimurium*, both with the addition of the S9 fraction and without the addition of this metabolization system, and for the reverse mutation test with the bacterium *Escherichia coli*. No increase in the sister chromatid exchange or chromosome aberrations in CHO cells and human lymphocytes exposed (*in vitro*) to toluene was reported, even at concentrations that inhibited the cell growth.

Inhalation of 1500 mg/L of toluene by male Wister rats induced oxidative damages in the lung, kidneys and liver, observed by the use of the immunoreactive marker 8-hydroxy-2'-deoxyguanosine (Tokunaga et al., 2003).

According to McGregor (1994), *in vitro* tests indicate that toluene is not genotoxic and *in vivo* evaluations carried out with rodents and humans are not conclusive regarding the genotoxicity of this compound. This same author suggests that exposures to toluene do not result in genetic damages in somatic cells. According to Hammer (2002), the genotoxic effect of toluene is still subject to much investigation.

As for the carcinogenic effect, the International Agency for Research on Cancer (IARC) considered toluene as belonging to Group 3 (not classified regarding its carcinogenicity), due to the absence of adequate data for such classification (WHO, 2011).

Biological Effects of Ethylbenzene

Ethylbenzene was classified by IARC (2000) as a possible carcinogenic substance to humans, due to significant results of carcinogenicity registered for animals.

This chemical passively cross the lipid bilayer of the cell membrane and can induce alterations both in the structure and in the integrity of the biological membranes. The changes in the integrity of the membrane can affect, particularly, the functions of the biological barrier, energy transduction, besides altering the protein matrix (Sikkema et al., 1995). In studies with the gastropod *Bellamya aeruginosa*, Zheng et al. (2012) observed that ethylbenzene was able to promote alterations in the retraction response of this mollusc and that such behavioural alteration was related with disturbances in the membrane caused by this substance.

Aromatic hydrocarbons, such as ethylbenzene, become metabolized when dissolved in an aqueous solution (Sikkema et al., 1995). Its preferential biotransformation occurs by hydroxylation catalyzed by enzymes of the hepatic microsomal fraction, with subsequent conjugation of its metabolites (Pedroso, et al., 2002). In man, the main metabolites formed by ethylbenzene are the mandelic and phenylglyoxylic acids (ATSDR, 1999).

Genotoxicity studies of ethylbenzene are controversial and further investigations are needed on this subject. Studies have been conducted on the potential of ethylbenzene in inducing chromosome aberrations and sister chromatid exchanges in CHO (NTP, 1986; 1996). In this study, it was not observed mutagenic response for the cells exposed to doses of 75, 100, or 125 mg/L. However, ethylbenzene was considered mutagenic in the concentration of 80 mg/L in assays conducted with mouse lymphoma (without metabolic activation) and lethal for cells of mouse lymphoma at the concentration of 100 mg/L (Mcgregor et al., 1988). The positive results suggest that ethylbenzene can cause an increase in the genotoxic potential in humans (ATSDR, 1999).

Zheng et al. (2012) also observed, by the comet assay in *B. aeruginosa*, a significative induction of DNA damages, when these organisms were exposed to concentrations of 45, 100, 450, 1000 μg/L of ethylbenzene, for exposure periods of 7, 14 and 21 days. The authors also reported a positive dose-response and time-response relationship.

Among the studies with ethylbenzene, *in vitro* assays with *S. typhimurium* (Florin et al., 1980; Dean, 1985; NTP 1986, 1996), *E. coli* (Dean, 1985) and *Saccharomyces cerevisiae* (Nestmann; Lee, 1983; Dean, 1985) were performed. These data show that ethylbenzene is not mutagenic for bacteria or yeast, both in the presence and absence of metabolic activity. A review conducted by Henderson et al. (2007) reports that ethylbenzene was not mutagenic for bacteria, did not induce mutations in yeast neither sister chromatid exchanges in CHO cells. This chemical was not considered clastogenic for CHO cells and cell lines from rat liver, but increased the formation of micronuclei in SHE cells (Syrian Hamster Embryo), when they were used in *in vitro* tests. Still according to the same authors, no evidence of genotoxicity was observed in humans exposed to high levels of ethylbenzene.

Due to the lack of genotoxicity observed for most of the assays developed with ethylbenzene, Zhang et al. (2010) investigated another possible mechanism of induction of renal tumours in Sprague-Dawley rats subjected by inhalation, for thirteen days, to 3 different concentrations of ethylbenzene (100, 1000 and 1500 mg/L). The authors concluded that this chemical is toxic to the kidney, inducing apoptosis and that the activation of the mitochondrial via may be one of the mechanisms involved in the carcinogenicity of ethylbenzene.

In relation to the embryotoxicity and teratogenicity, no lethal effect or malformations were observed in foetuses of pregnant Sprague-Dawley rats exposed for 15 days (from the 6th day until the 20th day of gestation) to the concentrations of 0, 250 and 1000 mg/L of ethylbenzene (Saillenfait et al., 2006). However, it was observed a significant decrease in the body weight for the highest concentration tested, indicating that this chemical can be harmful to the foetal development.

Yan and Zhou (2011), when conducting assays with the aquatic macrophyte *Hydrilla verticillata*, concluded that ethylbenzene is able to promote alterations in the lipid peroxidation and in the activity of antioxidant enzymes (superoxide dismutase and guaiacol peroxidase) in plants. Moreover, it was observed a dose dependent decrease in the amount of chlorophyll a and b produced by this organism, when it was submitted to different concentrations of this chemical (0.1, 1.0, 5.0, 10.0, 50.0, 100.0 mg/L).

Biological Effects of Xylene

Xylene is a compound with the characteristic of rapid absorption by inhalation. Both in its liquid and in the vapour form, xylene is slowly absorbed by the skin. However, after absorption, this chemical is rapidly distributed throughout the body, preferably to the adipose tissue. About 95% of xylene is metabolized by the liver and 70 to 80% of the metabolites formed are excreted by urine, in a period of 24 hours (Langman, 1994; WHO, 2011).

When xylene is inhaled, it can cause, depending on the degree of exposure, haemorrhage, inflammation and pulmonary oedema. However, these symptoms are hardly observed, since xylenes are irritating and present recognizable odour in concentrations much lower than those necessary to produce oedema. Aspiration of the liquid into the lungs can result in chemical pneumonia, which can be fatal (Meditext, 2000). Acute intoxications by xylene may cause transient renal alterations with elevation of the quantity of urea in the blood and decrease of urinary creatine (OPS, 1996). Neurotoxic action of xylene is manifested by cognitive dysfunction, increase in the levels of hypothalamic catecholamines, until a motor hyperactivity and convulsions (ATSDR, 1995).

Xylene is biotransformed by oxidation of its methyl grouping, forming the acid compounds ortho, meta or para-toluic [referring to each one of its three isomers: ortho (o-), meta (m-) and para (p-) xylene] and, after its conjugation with glycine, originates its main metabolite, the methylhippuric acid (Leite, 1996).

In rats, the exposure to xylene resulted in embryotoxic and teratogenic effects, affecting brain, liver, lungs and heart of these organisms (NLM, 1988). Exposure of pregnant Sprague–Dawley rats to o-, m- and p-xylene, from the sixth to the twentieth day of gestation, caused foetal toxicity evidenced by significative reduction of the body weight of the foetus in the two highest concentrations tested (1000 and 2000 mg/L) for m- and p-xylene. For o-xylene, this effect was already observed at the concentration of 500 mg/L (Saillenfait et al., 2003). According to Johnson (1986), exposure of artificial embryos of the cnidarian *Hydra attenuata* to the isomers of xylene and the mixture of these isomers did not induce embryotoxicity. However, toxic effects were observed for the embryos, but only for the same concentrations that also induced toxicity in the adult organisms of this species.

Negative results for xylene were observed in mutagenicity assays with *Salmonella* and L5179Y mouse lymphoma cells and in assays of chromosome damages with bone marrow cells exposed to dosages of xylene (NLM, 1988). However, xylene frequently occurs associated with other aromatic compounds, and this mixture is possibly more genotoxic than when it appears isolated (Irwin et al., 1997). According to Dean (1985), xylene can be considered as non mutagenic due to the conclusively negative results obtained with this compound with bacteria, mammalian cell culture and with animals *in vivo*. According to the World Health Organization (WHO, 1986; 1997; 2011), xylene does not present genotoxic or carcinogenic effects for humans and laboratory animals.

Toxicity studies with *Daphnia magna* show that ethylbenzene and m-xylene are the most toxic among the BTEX, when their concentrations in mass/volume are analyzed, while benzene is potentially more toxic, since it has the lowest LC50 among the compounds of BTEX (Irwin et al., 1997).

According to the World Health Organization (WHO, 1986; 1997; 2011), xylene does not present genotoxic or carcinogenic effects for human and laboratory animals.

Regarding the effect of xylene to plants, it was observed in *H. verticillata* that the compound induced alterations in the production of photosynthetic pigments, alterations in the lipid peroxidation and in the activity of antioxidant enzymes (Yan; Zhou, 2011).

Microbial Biodegration to Remediate BTEX-Contaminated Groundwater

An alternative for the removal of BTEX in aquifers contaminated with this mixture is the bioremediation process that, according to Barthe (1986) and Bouwer et al. (1994), is an attractive, safe and effective method against the conventional physico-chemical techniques that are employed in the treatment of soils and waters contaminated by petroleum derivatives.

The conventional physical treatments are not entirely advantageous since besides withdrawing the contaminants of the environment without destroying or chemically transform them, generating an accumulation of toxic residues, have a fairly high cost. The biological processes of bioremediation are considered a clean and promising technology, mainly due to their simplicity, low cost and high efficiency, when compared to the other alternatives (Alexander, 1994).

According to Mariano (2006), bioremediation can be considered as a promising technology to treat contaminated sites by the use of biological agents capable of modifying or decomposing target pollutants. The principle of this technique consists on the removal of pollutants from the natural environment and/or their conversion into less harmful products, using, for this, communities of endogenous microorganisms from the contaminated environment (Bamforth; Singleton, 2005). As a result, there is a mineralization of these pollutants, so as to transform them into smaller inorganic compounds, such as carbonic gas, water, inorganic forms of N, P and S and cell components (Alexander, 1994; Eweis et al., 1998).

The bioremediation technology by microorganisms is implemented in environments impacted by favouring the growth and reproduction of microorganisms, by providing an electron acceptor (oxygen or nitrate), nutrients (nitrogen and phosphorous) and a source of energy (carbon), which serves to accelerate the biodegradation process. As the petroleum and its derivatives are basically constituted by hydrocarbons, the biodegradation of these products is given by the use of these organic molecules as source of energy (Pedrozo et al., 2002).

Biodegradation generally occurs by a consortium of different species of microorganisms capable of metabolizing the products and by-products of biodegradation, in different stages (RÉGIS, 2000). The biodegrading agents can belong to different taxonomic groups such as algae, bacteria, fungi and protozoa (Ghiorse; Wilson, 1988; Melo; Azevedo, 1997). The microbial community involved in the degradation of environmental pollutants can be divided into: primary and secondary microorganisms. Primary microorganisms are those that are capable of metabolizing the main substrate supplied to the system, while the secondary microorganisms use, as source of energy, the products released by the primary microorganisms (Bull; Slater, 1982 apud Grady, 1985).

In relation to the organic compounds, most of them can be metabolized by microorganism, since the physico-chemical conditions of the environment are adequate for the needs of these microorganisms. Moreover, it should be also considered the proximity of these microorganisms to the target compound of the biodegradation, the availability of the compound and the participation of the compound in any metabolic via of the microorganisms (Régis, 2000; Silva et al., 2002; Inazaki, 2003).

For the biodegradation of hydrocarbons, it is essential that there is an oxi-reduction reaction, in which the hydrocarbon is oxidized (electron donor) and an electron acceptor is reduced. There are different compounds that can act as electron acceptors, among them the oxygen (O_2), nitrate (NO^{3-}), and oxides of Fe (III) and sulphate (SO_4^{-2}). Besides the electron acceptors, other variables can be related with the biological processes, such as pH and redox potential of the substance (Kao; Wang, 2000; Röling; Versevel, 2002; Hunkeler et al., 2002, Cheon et al., 2004).

Chiang et al. (1989) in a study of the aerobic biodegradation of BTEX in a shallow sand aquifer reinforced the idea that the quantity of BTEX in groundwater is inversely proportional to the availability of dissolved oxygen. The authors observed that when the concentration of dissolved oxygen was higher than 0.9 mg/L, BTEX was not detected due to their use in the biodegradation process.

Among the possible bioremediation techniques it is included: the use of autochtone microorganisms, i.e., from the own site, without any interference of active remediation technologies (intrinsic bioremediation or natural attenuation); addition of stimulating agents such as nutrients, oxygen and biosurfactants (biostimulation) and inoculation of enriched microbial consortia (bioaugmentation) (Bento et al., 2003).

Natural attenuation appears as an interesting process for treatment of groundwaters contaminated with hydrocarbons, mainly due to its low cost and by being a technique that requires a minimal intervention of the impacted environment (Bhupathiraju et al., 2002). The main biotransformation mechanisms of the organic compounds in groundwaters occur in the biofilms. Biofilms are formed by bacteria and extracellular polymers that adhere into the surface of a substrate (intermediate layer and interface water-sediment), obtaining energy and nutrients by the flow of groundwater (Bitton; Gerba, 1984). The bioremediation strategies implies in providing appropriate conditions of water, air and supplementation of nutrients for microorganisms, in order to promote the metabolization of the contaminants (Bamforth; Singleton, 2005).

The chemical structure of the organic pollutants has a profound influence in the ability of the microorganisms of metabolizing these molecules, especially with respect to the rates and extension of biodegradation. Some organic compounds are rapidly biodegraded (such as hydrocarbons with low to medium molecular weight and alcohols), while others are recalcitrant (such as the more complex chemical compounds and, mainly, the halogenated hydrocarbons) (Mariano, 2006). Generally the branched and polynucleated molecules are more difficult to degrade than the monoaromatic molecules or compounds of single strands compounds (Alexander, 1965 apud Atlas, 1997).

The ability of microbial degradation of BTEX components was first reported by Stormer in 1908, who observed that the bacterium *Bacillus hexacarbovorum* was able to use toluene and xylene as source of carbon (Zobell, 1946). In 1914, Wagner described that the bacterium *Bacterium benzoli* was also able to use BTX as energy (Zobell, 1946). In 1928, Gray and Thornton were the first to show that the microorganisms naturally present in the soil was able degrade BTEX. The authors identified that from the 245 species isolated from the soil, 146 of them were able to degrade hydrocarbons (Gibson; Subramaniam, 1984; Corseuil; Alvarez, 1996). From these pioneer studies, the bioremediation process by microorganism has been the focus of countless studies aiming to obtain a safer and more effective methodology to decontaminate sites impacted with BTEX, in particular groundwater.

Approximately 250 bacteria were isolated from an aquifer contaminated with 20,000 to 40,000 litres of unleaded gasoline located near the U.S. Naval Weapons Station in Seal Beach, California (Ridgway et al., 1990). Among the bacteria identified, different strains (with distinct catabolic pathways) of the species *Pseudomonas aeruginosa*, *P. putida*, *P. maltophillia* and bacteria of the genus *Micrococcus* were considered as capable of degrading the four constituents of BTEX.

Mazzeo et al. (2010) promoted the aerobic degradation of BTEX using an inoculum obtained from a petroleum refinery effluent. Although this inoculum was constituted by 5 different types of bacteria, *P. putida* was the only species considered responsible for the total degradation of BTEX. According to the authors, the biodegradation success was mainly due to the selection of microorganisms through enrichment, which made the microorganisms more adapted to the target pollutants. Villatoro-Monzón et al. (2003) also reported that environments that have already suffered contaminations by aromatic hydrocarbons can be source of potentially degrading microorganisms of BTEX since they are already naturally pre-selected.

Lee et al. (1994) modified the *tol* and *tod* metabolic pathway of the bacterium *P. putida*, creating the TB105 strain, which is able to mineralize, simultaneously, benzene, toluene and p-xylene, without generating any intermediate metabolite during the degradation. The authors believe that the construction of microorganisms genetically modified can be of great importance in the bioremediation of environmental pollutants.

Lee et al. (2012), after sequencing the genome of the bacterium *Pseudoxanthomonas spadix* BD-a59, found out that this bacterium has a quite high number of genes of enzymes related with the metabolism of recalcitrant organic compounds such as BTEX. It was found twelve monooxygenase and 28 dioxygenase genes, which are related with the total degradation of the six constituents of BTEX (benzene, toluene, ethylbenzene, m-xylene, p-xylene and o-xylene).

Considering that the amount of oxygen available in the groundwater can be a limiting factor and that some microorganisms are able to substitute oxygen for other electron acceptors in hypoxia conditions or anaerobiosis, Shim et al. (2005) proposed the use of a coculture of *P. putida* and *P. fluorescens* to degrade BTEX. During the experiment the authors provided hydrogen peroxide as an additional source of oxygen; BTEX as only source of carbon and nitrate as electron acceptor, for *P. fluorescens*, under hypoxia conditions. The authors concluded that the coculture was able to completely mineralize benzene and toluene, and may be a promising tool in the treatment of industrial effluents and groundwater contaminated with BTEX.

Dou et al. (2008) also evaluated the potentiality of microorganisms to anaerobically degrade BTEX under nitrate reducing conditions. The authors used an enriched mixed culture composed by *P. aureofaciens*, *Microbacterium Lactuim* and *B. cereus* to investigate the interactions between the constituents of BTEX during their anaerobic biodegradation. This study concluded that the interactions are quite complex and that they occur in three ways: (1) presence of a constituent of BTEX (e.g. toluene) stimulates the degradation of the others; (2) presence of a constituent of BTEX (e.g. benzene, p-xylene) inhibits the degradation of the others; (3) presence of low concentrations of a constituent of BTEX stimulates the degradation of the others, whereas high concentrations have an opposite effect (e.g. presence of toluene and ethylbenzene in relation to the degradation of m-xylene and p-xylene).

From the use of a mixed culture constituted by *Pseudomonas* sp, *Yarroia* sp, *Acinetobacter* sp, *Corynebacterium* sp, and *Sphingomonas* sp, Jo et al. (2008) observed a different response for the interactions that occur between the components of BTEX during the biodegradation process, when they constitute a quaternary mixture. These researchers found that the increase in the concentration of xylene in the mixture favoured the removal of the other constituents, although an increase in the concentration of benzene has also entailed in a decrease of the biodegradation efficiency.

According to Villatoro-Monzón (2003), anaerobic biodegradation of BTEX occurs more efficiently when Mn (IV) is used as electron acceptor, when compared with Fe (III), mainly in relation to the degradation of benzene. However, as these metals are easily found in the environment, biostimulation using manganese or iron represents a promising and low cost alternative.

The process of *in situ* biostimulation was performed by Silva and Corseuil (2012) to favour the biodegradation of BTEX in a gasohol-contaminated aquifer located in the State of Santa Catarina (Brazil). In this study, it was observed that the injection of nitrate contributed for a further acceleration of the process, when compared to the natural attenuation. In relation to the local microbial composition, the addition of nitrate did not favour the development of anaerobic microorganisms, but provided the growth of aerobic microorganisms by contributing for the decrease of biochemical oxygen demand.

Air injection to promote oxygenation and subsequent aerobic biodegradation of groundwater contaminated by the leakage of an oil pipeline at Southern Taiwan resulted in an increase in the removal efficiency of BTEX from 92% to 99%, showing that this type of biostimulation also contributes to accelerate the removal process of BTEX by biodegradation (Chen et al., 2010).

According to Prenafeta-Boldú et al. (2002), the potential of biodegrading BTEX by fungal species should not be ignored. Their affirmation was based in the capacity of the deuteromycete fungus *Cladophialophora* sp. strain T1 (= ATCC MYA-2335) oxidize the compounds toluene, ethylbenzene and xylenes in the alkyl radical by the monooxygenase enzyme, using toluene and ethylbenzene as source of carbon and energy. According to

García-Peña (2008), the fungus *Paecilomyces variotii* CBS115145 has the potential of degrading BTEX, although benzene and xylenes were partially degraded. The authors also concluded that the binary or multiple combination of the constituents of BTEX reduces the degradation rates of these compounds, suggesting an inhibition of the catabolic pathways of the fungus by the presence of structurally related molecules. Rene et al. (2012) observed that the fungus *Exophiala* sp is also capable of biodegrading BTEX.

In a study performed with the fungi *Cladophialophora* sp and *Cladosporium* sp, isolated from sites contaminated with gasoline, it was found that ethylbenzene was easily degraded by both fungi. However, none were able to degrade benzene (Nikolova; Nenov, 2005). According to the authors, better results of bioremediation were obtained with *Cladophialophora* sp, except only for p-xylene.

The process of natural attenuation in contaminated groundwaters is an important method to decrease the concentration and retain the contaminant plume (Chen et al., 2010). According to the same authors, factors such as consumption of dissolved oxygen, nitrate and sulphate; production of iron ions, sulphite, methane and CO_2; decrease of pH of the local source of pollution and increase of pH in the iron reducing area; and preferential reduction of the concentration of some constituents of BTEX along the transport route are evidences that the biodegradation of BTEX is occurring in groundwaters that are suffering intrinsic bioremediation.

According to Choi and Lee (2011), the hydrogeological characteristics interfere directly in the natural attenuation process of BTEX-contaminated groundwater. Studying an aquifer located at a military installation in Korea, the authors observed that the high water level, permeable upper layer and constant rains favoured the natural attenuation of BTEX. Biodegradation, assessed by the observation of alterations in the concentrations of the electron acceptors and donors, was quantified at a rate between 20.52 and 33.67 mg/L. Denitrification (61.32%) was the process that most contribute to the occurrence of biodegradation. However, dilution and dispersion of this pollutant also provided its natural attenuation. The authors concluded that, although natural attenuation can be an effective alternative for the remediation of low levels of this contaminant, this process can be very slow and, therefore, need additional procedures associated with it.

Davis et al. (1999) evaluated, during 5 years, the natural attenuation process of groundwater contaminated with fuel derived from rupture of a subterranean tank located in the metropolitan area of Perth, Australia. The authors observed that there was a preferential degradation of TEX, and the sulphate was the main electron acceptor.

An effective known treatment for use in contaminated groundwaters is the technique of permeable reactive barrier. In this methodology, the contaminated groundwater flows through a reactive material, where the contaminants are removed by physical, chemical and biological processes (Gavaskar, 1999). The permeable reactive barrier was used by Yeh et al. (2010) to investigate the relationship between the efficiency of the BTEX degradation and the distribution of the microbial community in the system. According to the authors, benzene and toluene are inhibitors of the microbiological activity, when present in concentrations above 80 mg/L. Moreover, effects of shock-loading can be harmful to the microbial community, leading to a decrease of the biomass and a consequent decrease in the removal of BTEX. The authors suggest that permeable reactive barrier systems can be useful in the *in situ* remediation of BTEX-contaminated groundwater. Lin et al. (2010) used an oxygen-releasing reactive barrier to treat a BTEX-contaminated groundwater in lab-scale. The culture of microorganisms used was constituted by *Pseudomonas* sp (DQ211692), *Pseudomonas* sp (AF065166), *P. putida* (AY686638) and Pseudomonas sp (DQ124297). A biostimulation was performed by the addition of oxygen-releasing compounds (e.g. CaO_2, KH_2PO_4, K_2HPO_4, $NaNO_3$), followed by a bioaugmentation. The results showed that both biostimulation and bioaugmentation contribute to accelerate the microbial activity and degradation of BTEX. However, the efficiency of this biodegradation was higher in the system that contained nitrogen, since this nutrient increases the effectiveness of degrading bacteria both in aerobic and anaerobic conditions. The authors also observed that factors such as dissolved oxygen, chemical oxygen demand (COD), bacteria densities and the microbial community structure must be considered, since they contribute for the efficiency of oxygen-releasing reactive barrier in the degradation of BTEX. Xin et al. (2012) combined the system oxygen-releasing permeable reactive barrier with the bioaugmentation technology using the species *Mycobacterium* sp CHXY119 and *Pseudomonas* sp YATO411 (in the form of immobilized beads) to remediate high concentration BTEX-contaminated groundwater (about 100 mg/L). The system used promoted a removal of almost 100% of the BTEX present, proving its great potential for practical applications in the remediation of groundwaters with high concentrations of these contaminants.

Effectiveness of the Bioremediation Process of BTEX by the Use of Bioassays

For a bioremediation be considered efficient, the product resulting from the decomposition should be constituted only by water and carbonic gas. However, several assays report that microbial activities can transform non toxic or with low toxicity compounds into potent toxic and in many cases products resulting from the bioremediation are recognizably carcinogenic (Alexander, 1999).

Analyzing soils contaminated by petroleum derivatives submitted to the bioremediation process, Alexander et al. (2002) observed a significative increase of the genotoxicity and suggested that the microorganism responsible for the degradation of the compounds were transforming the chemicals into new substances with evident genotoxic potentials. Phillips et al. (2000) reported in their studies conducted with soil that, although the bioremediation treatment reduced the amount of total petroleum hydrocarbons, it induced an increase in the toxicity, as observed in the tests of seed germination and toxicity with earthworm, indicating that intermediate toxic metabolites could be formed during the biodegradation. Microbial activities, in which non toxic compounds or with low toxicity are converted to potent toxic, have been reported for several compounds and some products are, recognizably, carcinogenic (Alexander, 1999). An increase in the genotoxicity was also observed by Randerath et al. (1999), when the authors analyzed, in laboratory, soils contaminated with diesel submitted to bioremediation.

The evaluation of the effectiveness of the bioremediation process is generally made only by chemical analyses. However, there is the eminent need to associate to these chemical evaluations other methodologies that help in the detection and quantification of molecular transformations, resulted from bioremediation processes. According to Maila e Cloete (2005), the use of biological tests as bioindicators of the removal of organic pollutants should be used as a complement to the existing traditional tools, and it is possible to understand the effect of the technological treatment and the environmental conditions in the responses presented by the bioindicators. Plaza et al. (2005) cited that the use of bioassays combined with chemical analyses present more reliable results for the analysis of genotoxic risks. This combination of assays, according to the authors, is indicated to evaluate the effectiveness of bioremediation processes. According to Dawson et al. (2007), the type of contaminant, the matrix where the contaminant is found and the remediation strategy adopted interfere directly in the choice of adequate bioassays to evaluate the progress of the bioremediation.

In the case of BTEX, different methodologies, including biological tests, can detect and estimate the transformations of benzene, toluene, ethylbenzene and xylene and, thus, evaluate the feasibility of the bioremediation process in the treatment of aquifers contaminated with these substances (Reusser et al., 2002).

Luminescence-based bacterial biosensors have been frequently used in studies of environmental contaminations and in the evaluation of strategies of BTEX bioremediation (Sousa et al., 1998; Dawson et al., 2008; Xin et al. 2013). According to Bhattacharyya et al. (2005), biosensors are considered early warning tools of contaminants and are cost-effective and easy to perform and interpret. Light-emitting bacterial biosensors can be naturally luminescent or genetically modified by the insertion of a lux gene (Sousa et al., 1998; Bhattacharyya et al., 2005). Such bioassays can be divided into two categories: those able to monitor the presence of a specific contaminant (specifically induced biosensor), in which the lux genes bind to a promoter belonging to the catabolic pathways of the contaminants, so that the luminescence is only produced in their presence; and those that respond to the amount of contaminant present in the sample (constitutively expressed biosensor), in which the lux genes are constitutively expressed, so that the luminescence is the measure of the metabolic activity of the cell, and is used as toxicity assay (Bundy et al. 2001; 2002).

A lux-based bacterial biosensor technology was used to monitor the degradation of BTEX of aqueous extracts of soils impacted with this contaminant (Dawson et al., 2008). The biosensor *P. putida* F1 Tn5, used to measure the toxicity of BTEX, showed an increase of the toxicity after the initiation of the bioremediation, possibly due to the accumulation of intermediate metabolites, derived from the partial oxidation. In the process, these metabolites were completely degraded, evidenced by the increase in the bioluminescence, indicative of the reduction of the toxicity in the samples. Another biosensor used to verify the availability of BTEX in the aqueous extracts was *P. putida* TVA8. As the induction of luminescence is proportional to the induction of the *tod* promoter gene, related to the degradation of hydrocarbons, it was possible, in this case, to observe a coincidence between the luminous response and the decrease in the concentration of BTEX, after 12 days of bioremediation. The authors suggested that the use of biosensors favours the understanding of the biodegradation process, by detecting the formation of potentially toxic intermediates and the alterations in the availability of the target contaminant.

Sousa et al. (1998) used the bioluminescence-based bacterial biosensor (*P. fluorescens* 10586s pUCD607) to detect the toxicity of other pollutants that could limit the bioremediation of BTEX, such as suspended solids, nonvolatile organopollutants and inorganic compounds, such as metals. The use of the simple techniques such as filtration, volatilization and incineration in muffle, enables the identification of the presence of interferings in groundwaters samples, allowing the authors to determine the contribution of each of the interfering in the toxicity of the samples. According to the authors, the use of this technology can provide more satisfactory results for the biodegradation of BTEX, since the presence of interferings can inhibit the microbial activity and make the bioremediation process unfeasible.

Xin et al. (2013) used the natural luminescent marine bacterium *Vibrio fischeri* to assess possible alterations in the toxicity of BTEX-contaminated groundwater, after 35 days of bioremediation, using the technology of oxygen-releasing permeable reactive barrier. The dilution value for 50% of the luminescent inhibition (EC50) dropped from 38.6 to 3.4. The authors believe that, although it was not obtained a total reduction of the toxicity, there was a decrease of 91.2% of the toxic potential of the studied groundwaters, which shows a great efficiency of the treatment used.

The microcrustacean *Ceriodaphnia dubia* (cladoceran) is internationally recognized as an excellent bioindicator of aquatic contaminants, due to its high sensitivity to a wide variety of contaminants, short reproductive cycle and possibility of evaluating different endpoints, including lethality, growth, reproduction and mobility (Gonzalez et al., 2009). Gersberg et al. (1995) used the test organism *C. dubia* to monitor the reduction of the toxicity of bioremediated groundwaters of an aquifer contaminated with gasoline in San Diego (California, EUA). Although the chemical analyses have showed a decrease exceeding 80% of the concentration of BTEX, reaching 99% in one of the sites, the acute toxicity test of 24-h showed an increase of the LC50 from 12-15% to 56%. In one site of the same aquifer not contaminated with BTEX the value of LC50 was of 74%, suggesting that the other contaminants present in the aquifer, already present above the contamination plume, induced toxicity in *C. dubia*. Carroquino et al. (1992) had already found similar results in the same study area and using the same test organism. However, the authors observed that BTX contributed to an increase in the toxicity observed in the area not contaminated of the aquifer. This study also showed that the toxicity assay with *C. dubia* was an excellent tool for the assessment of the reduction of the toxicity of BTX-contaminated groundwaters, after the bioremediation process, besides enabling the evaluation of which type of bioremediation process can be more efficient.

The acute toxicity test of 48-h performed by Nunes-Halldorson et al. (2004), using the species *C. dubia* showed that, although the biodegradation of BT have been efficient (98%), the toxicity of the samples, despite reduced, remained high after biodegradation. The authors suggested that this residual toxicity is due to the formation of toxic intermediates, such as catechol. However, as the degradation period was of only 1 week, it is possible that if there was an increase in this treatment period, the metabolites could have also been consumed. As this bioindicator is not specific to determine the toxicity of BTEX, but presents a high sensitivity to wide diversity of chemicals, the presence of other contaminants of the study site could have interfered in the evaluation of the efficiency of the BTEX degradation. However, in environmental recovery studies, the detection of the toxicity by the test serves as an alert that the other contaminants can be also impacting the area.

Among higher plants, *A. cepa* has been indicated as an efficient test organism for studies of basic mechanisms of action and determination of the effects of some chemicals (Fiskejö, 1985; Bushra Ateeq et al., 2002), due to the characteristics in its proliferation kinetics, fast growth of the roots, large number of cells in division, high tolerance to different cultivation conditions, availability throughout the year, easy handling and for having chromosomes in a reduced number (2n=16) and of large size (Quinzani-Jordão, 1978; Rank; Nielsen, 1998; Patra; Sharma, 2002; Matsumoto et al., 2006), besides being easily stained and observed (Kurás et al., 2006). The analysis of meristematic cells of *A. cepa* enables the monitoring of several *endpoints*, such as induction of chromosome and chromatidic aberrations, sister chromatid exchanges, alterations in the mitotic index and formation of micronuclei (Migid et al., 2007).

The application of the chromosome aberrations and micronucleus test in meristematic cells of *A. cepa*, allowed concluding that the bacterial biodegradation of aqueous solutions of BTEX was very efficient, due to the decrease observed in the genotoxic effects, after the biodegradation process (Mazzeo et al., 2011). From the results, the authors affirmed that no toxic intermediate metabolite was formed during the biodegradation of this mixture. Moreover, the same authors recognized the species *A. cepa* as a good indicator to prove the efficiency of bioremediation processes with these hydrocarbons.

The use of *in vitro* tests, besides not bumping into ethical factors, leads to a decrease in the number of animals used in experiments, a reduction in the operational costs and infra-structure (OTA, 1986). *In vitro* tests, with mammal cells, can be used in studies to evaluate the mutagenicity and genotoxicity of environmental chemicals (Cardozo et al., 2006). According to Rogero et al. (2003), cell culture presents advantages over the *in vivo* methods due to the possibility of limiting the number of experimental variations, obtaining significant data more easily and for having shorter periods of experimentation.

Rat hepatoma tissue culture (HTC) cells were used as test systems in the evaluation of the reduction of the toxic effects of BTEX after biodegradation (Mazzeo et al., 2013). With the use of the comet assay to detect the induction of damages in the DNA, a significant decrease of these damages was observed for all the biodegraded concentrations tested. In relation to the presence of micronuclei, it was observed a significative frequency for the highest concentration tested (near the solubility of each one of the components of the BTEX in water) both before and after the biodegradation process. However, this significant effect after the biodegradation refers to according to the authors the incomplete degradation of all the BTEX present, which reached only 64% of their total value. Nevertheless, the authors suggest that the biological assays used are quite efficient to be used in the assessment of the effectiveness of the bioremediation process of BTEX.

Conclusion

Due to the high use of BTEX as solvent and raw material in the fabrication of pesticides, plastics, adhesives, synthetic fibres, paints, and pharmaceutical products and to its high concentration in fossil fuels, BTEX is a constant pollutant in the environment. The frequent presence of this contaminant is quite worrying, since BTEX has a proven toxic effect. Although each one of its components presents a specific action, there is still a potential synergistic action for the mixture of these hydrocarbons.

The concerns related to contaminations by hydrocarbons have stimulated the development of several methodologies that enable the removal of these compounds from contaminated sites. Among the possible clean-up technologies, the bioremediation process is one of the most studied and promising. A wide variety of microorganisms can act in the biodegradation of hydrocarbons, among them the monoaromatics such as BTEX. Although some fungal species have the ability to degrade this pollutant, bacteria seem to contribute in a more effective form to this, both due to the higher diversity of degrading species and for the better results observed.

Chemical interactions between the components of BTEX are complex, and the presence of a component can favour or difficult the degradation of another, helping or impairing the total elimination of the contamination. Moreover, the concentration of each one of the constituents of BTEX can be a limiting factor for its degradation, by inhibiting the growth of microorganisms. Among the components of the BTEX complex, benzene seems to be the most limiting constituent. Aerobic biodegradation is the preferential and the most effective via in the degradation of BTEX, but in environments with reducing conditions, the biostimulation by the addition of electron acceptors can be also considered quite satisfactory.

According to the data reported in the literature, the use of biological assays contributes in a very effective form to monitor this process, allowing the identification of toxic intermediate metabolites, detection of non target contaminants that can limit the process as well as evaluate the possible reduction of the toxicity of the target pollutant.

Although BTEX is a frequent pollutant of groundwater, different types of technologies are already available to promote the clean-up of these areas. By the reports found in the literature, it can be concluded that the bioremediation process is efficient in the degradation of BTEX, both in relation to the chemical part (absence of the compound at the end of the process) and in relation to the biological part (complete mineralization of the contaminant without the presence of toxic metabolites). The toxicity observed in some studies, after the bioremediation process, seems to be more related to the presence of BTEX in the medium, not consumed in its totality, or the presence of metabolites that can still be consumed, than to the formation of toxic intermediate metabolites that cannot be metabolized.

REFERENCES

Aitio, K., Wada, H., (1993). Behavioral approaches to toluene intoxication. *Environ. Res.*, 62, 53-62.

Alexander, M. *Biodegradation and Bioremediation*. Michigan: Academic Press, 1994. 302 p.

Alexander, M. *Biodegradation and Bioremediation.* San Diego: Academic Press, 1999, 453p.

Alexander, R. R., Tang, J., Alexander, M., (2002). Genotoxicity is unrelated to total concentration of priority carcinogenic polycyclic aromatic hydrocarbons in soils undergoing biological treatment. *J. Environ. Qual.*, 31, 150-154.

An, Y.-J., (2004). Toxicity of benzene, toluene, ethylbenzene, and xylene (BTEX) mixtures to *Sorghum bicolor* and *Cucumis sativus*. *Bull. Environ. Contam. Toxicol.*, 72, 1006-1011.

Angelini, S., Kumar, R., Bermejo, J. L., Maffei, F., Barbieri, A., Graziosi, F., Carbone, F., Cantelli-Forti, G., Violante, F. S., Hemminki, K., Hrelia, P., (2011). Exposure to low environmental levels of benzene: evaluation of micronucleus frequencies and S-phenylmercapturic acid excretion in relation to polymorphisms in genes encoding metabolic enzymes. *Mutat. Res.*, 719, 7-13.

Anneser, B., Einsiedl, F., Meckenstock, R. U., Richters, L., Wisotzky, F., Griebler, C., (2008). High-resolution monitoring of biogeochemical gradients in a tar oil-contaminated aquifer. *Appl. Geochem.*, 23, 1715-1730.

Atlas, R. M. Applicability of bioremediation to eastern European pollution problems. In: *Training Workshop of Ics-Unido on "Soil Environmental Assesment And Bioremediation Technologies"*, 1997.

ATSDR. Agency for Toxic Substances and Disease Registry. *Toxicological profile for xylenes.* Atlanta: US Department of Health and Human Services, 1993. 209p.

ATSDR. Agency for Toxic Substances and Disease Registry. *Toxicological profile for xylenes.* Atlanta: US Department of Health and Human Services, 1995.

ATSDR. Agency for Toxic Substances and Disease Registry. *Toxicological profile for ethylbenzene.* Atlanta: US Department of Health and Human Services, 1999. 238 p.

ATSDR. Agency for Toxic Substances and Disease Registry. *Interaction profile for: Benzene, Toluene, Ethylbenzene, and Xylenes (BTEX).* Atlanta: US Department of Health and Human Services, 2004, 137p.

Au, W. W., Ramanujam, V.M.S., Ward JR., J.B., Legator, M.S., (1991). Chromosome aberrations in lymphocytes of mice after sub-acute low-level inhalation exposure to benzene, *Mutat. Res.*, 260, 219-224.

Bamforth, S. M, Singleton, I., (2005). Bioremediation of polycyclic aromatic hydrocarbons: current knowledge and future directions. *J. Chem. Technol. Biotechnol.*, 80, 723-736.

Barreto, G. E. S., Oliveira, E. S., Villar, F. A. G. A., Santos, G. S., Lima, R. M. F., Costa, S. L., Costa, M. F. D., Egito, E. S. T., El-Bachá, R. S., (2007). Investigation of toxic factors affecting cells of rat brains exposed to 3-methylcatechol. *Braz. Arch. Biol. Technol.*, 50, 839-849.

Barthe, R., (1986). Biotechnology of petroleum pollutant biodegradation. *Fems Microbiol. Ecol.*, 12, 155-172.

Bento, F. M., Camargo, F. A. O., Okeke, B., (2003). Bioremediation of soil contaminated by diesel oil. *Braz. J. Microbiol.*, 34, 65-68.

Bhattacharyya, J., Read, D., Amos, S., Dooley, S., Killham, K., Paton, G.I., (2005). Biosensor-based diagnostics of contaminated groundwater: assessment and remediation strategy. *Environ. Pollut.*, 134, 485-492.

Bhupathiraju, V. K., Krauter, P., Holman, H. N., Conrad, M. E., Daley, P. F., Templeton, A. S., Hunt, J. R., Hernandez, M., Alvarez-Cohen, L., (2002). Assessment of in-situ bioremediation at a refinery waste-contaminated site and an aviation gasoline contaminated site. *Biodegrad.*, 13, 79-90.

Bird, M. G., Greim, H., Snyder, R., Rice, J.M., (2005). International symposium: Recent advances in benzene toxicity. *Chem.-Biol. Interact.*, 153-154, 1-5.

Bitton, G.; Gerba, C. P. *Ground water. Pollution Microbiology.* N.Y.: John Wiley & Sons., 1984, 377p.

Bollati, V., Baccarelli, A., Hou, L., Bonzini, M., Fustinoni, S., Cavallo, D., Byun, H.-M., Jiang, J., Marinelli, B., Pesatori, A. C., Bertazzi, P. A., Yang, A. S., (2007). Changes in DNA methylation patterns in subjects exposed to low-dose benzene. *Cancer Res.*, 67, 876-880.

Bouwer, E., Durant, N., Wilson, L., Zhang, W., Cunningham, A., (1994). Degradation of xenobiotic compounds in situ: capabilities and limits. *Fems Microbiol. Ecol.*, 15, 307-317.

Bundy, J. G., Campbell, C. D., Paton, G. I., (2001). Comparison of response of six different luminescent bacterial bioassays to bioremediation of five contrasting oils. *J. Environ. Monitor.*, 3, 404-410.

Bundy, J. G., Paton, G. I.; Campbell, C. D., (2002). Microbial communities in different soil types do not converge after diesel contamination. *J. Appl. Microbiol.*, 92, 276-288.

Bushra Ateeq, M., Abul Farah, M., Niamat Ali, W. A., (2002). Clastogenicity of pentachlorophenol, 2,4-D and butachlor evaluated by *Allium* root tip test. *Mutat. Res.*, 514, 105-113.

Cardozo, T. R., Rosa D. P., Feiden, I. R., Rocha, J. A. V., Oliveira, N. C. D., Pereira, T. S., Pastoriza, T. F., Marques, D. M., Lemos, C. T., Terra, N. R., Vargas, V. M. F., (2006) Genotoxicity and toxicity assessment in urban hydrographic basins. *Mutat. Res.*, 603, 83-96.

Carroquino, M. J., Gersberg, R. M., Dawsey, W. J., Bradley, M. D., (1992). Toxicity reduction associated with bioremediation of gasoline-contaminated groundwaters. *Bull. Environ. Contam. Toxicol.,* 49, 224-231.

Chen, H., Rupa, D. S., Tomar, R., Eastmond, D. A., (1994). Chromosomal loss and breakage in mouse bone marrow and spleen cells exposed to benzene *in vivo. Cancer Res.*, 54. 3533-3539.

Chen, C. S., Hseu, Y. C., Liang, S. H., Kuo, J.-Y., Chen, S.C., (2008). Assessment of genotoxicity of methyl-tert-butyl ether, benzene, toluene, ethylbenzene, and xylene to human lymphocytes using comet assay. *J. Hazard. Mater.*, 153, 351-356.

Chen, K.-F., Kao, C.-M., Chen, C.-W., Surampalli, R.Y., Lee, M.-S., (2010). Control of petroleum-hydrocarbon contaminated groundwater by intrinsic and enhanced bioremediation. *J. Environ. Sci.*, 22, 864-871.

Cheon, J. Y., Lee, J. Y., Lee, K. K., (2004). Characterization of the hydrogeologic environment at a petroleum hydrocarbon contaminated site in Korea. *Environ. Geol.*, 45, 869- 883.

Chiang, C. Y., Salanitro, J. P., Chai, E. Y., Colthart, J. D., Klein, C. L., (1989). Aerobic biodegradation of benzene, toluene and kylene in a sand aquifer: data analysis and computer modeling. *Ground Water Monit. R.*, 27, 823-834.

Choi, H.-M., Lee, J-Y., (2011). Groundwater contamination and natural attenuation capacity at a petroleum spilled facility in Korea. *J. Environ. Sci.*, 23, 1650-1659.

Corseuil, H. X., Alvarez, P. J. J., (1996). Natural bioremediation perspective for BTX contaminated groundwater in Brazil: effect of ethanol. *Water Sci. Technol.*, 34, 311-318.

Dabney, B. *REPROTEXT data base*, Vol.22, Denver: Micromedex Inc., 1994.

Davis, G. B., Barber, C., Power, T. R., Thierrin, J., Patterson, B. M., Rayner, J. L., Wu, Q., (1999). The variability and intrinsic remediation of a BTEX plume in anaerobic sulphate-rich groundwater. *J. Contam. Hydrol.,* 36, 265-290.

Dawson, J. J. C., Godsiffe, E. J., Thompson, I. P., Ralebitso-Senior, K., Killham, K. S., Paton, G. I., (2007). Application of biological indicators to assess recovery of hydrocarbon impacted soils. *Soil Biol. Biochem.*, 39, 164-177.

Dawson, J. J. C., Iroegbu, C. O., Maciel, H., Paton, G. I., (2008). Application of luminescent biosensors for monitoring the degradation and toxicity of BTEX compounds in soils. *J. Appl. Microbiol.*, 104, 141-151.

Dean, B. J., (1978). Genetic toxicology of benzene, toluene, xylenes and phenols. *Mutat. Res.*, 47, 75-97.

Dean, B. J., (1985). Recent findings on the genetic toxicology of benzene, toluene, xylenes and phenols. *Mutat. Res.*, 154, 153-181.

Dou, J., Liu, X., Hu, Z., (2008). Substrate interactions during anaerobic biodegradation of BTEX by the mixed cultures under nitrate reducing conditions. *J. Hazard. Mater.*, 158, 264-272.

Duarte-Davidson, R., Courage, C., Rushton, L., Levy, L., (2001). Benzene in the environment: an assessment of the potential risks to the health of the population, *Occup. Environ. Med.*, 58, 2-13.

EC. Environment Canada. *Priority substances list assessment report on toluene.* Quebec: EC, 1993. 26 p.

Esteve-Turrillas, F. A., Ly-Verdú, S., Pastor, A., Guardia, M., (2009). Development of a versatile, easy and rapid atmospheric monitor for benzene, toluene, ethylbenzene and xylenes determination in air. *J. Chromatog. A*, 1216, 8549-8556.

Eweis, J. B., Ergas, S. J., Chang, D. P. Y., Schroeder, E. D. *Bioremediation Principles.* WCB/McGraw-Hill, 1998, 530p.

Falcó, I. P., Moya, M. N. *Analysis of volatile organic compounds in water. In: NOLLET, L.M.L. (Ed.). Handbook of water analysis.* New York: CRC Press, p.599-666, 2007.

Fiskejö, G., (1985). The *Allium* test as a standart in environmental monitoring. *Hereditas*, 102, 99-112.

Florin, I., Rutberg, L., Curvall, M., (1980). Screening of tobacco smoke constituents for mutagenicity using the Ames test. *Toxicol.*, 18, 219-232.

García-Peña, I., Ortiz, I., Hernández, S., Revah, S., (2008). Biofiltration of BTEX by the fungus *Paecilomyces variotii. Int. Biodeter. Biodegr.*, 62, 442-447.

Gavaskar, A. R., (1999). Design and construction techniques for permeable reactive barriers. *J. Hazard. Mater.*, 68, 41-71.

Gersberg, R. M., Carroquino, M. J., Fisher, D. E., Dawsey, J., (1995). Biomonitoring of toxicity reduction during in situ bioremediation of monoaromatic compounds in groundwater. *Water Res.*, 29, 545-550.

Ghiorse, W. C., Wilson, J. L., (1988). Microbial ecology of the terrestrial subsurface. *Adv. Appl. Microbiol.*, 33, 107-172.

Gibson, D. T., Subramanian, V. Microbial degradation of aromatic hydrocarbons. In: GIBSON, D.T., (Ed.) *Microbial degradation of organic compounds.* New York: Marcel Dekker Inc; 1984. p.181-252.

Góes, R. C. S. *Toxicologia industrial: um guia prático para a prevenção e primeiros socorros.* Rio de Janeiro: Revinter, 1997.

Gonzalez, C.; Greenwood, R.; Quevauviller, P.P. *Rapid Chemical and Biological Techniques for Water Monitoring.* Wiley, 2009, 440p.

Grady, C. P. L., (1985). Biodegradation: its measurement and microbiological basis. *Biotechnol. Bioeng.*, 27, 660-674.

Green, M. *Academic American Encyclopedia*, Danbury: Grolier Electronic Publishing, 1988.

Hammer, K. D., (2002). Metabolite ratio of toluene-exposed rotogravure printing plant workers reflects individual mutagenic risk by sister chromatid exchanges. *Mutat. Res.* 519, 171-177.

Henderson, L., Brusick, D., Ratpan, F., Veenstra, G., (2007). A review of the genotoxicity of ethylbenzene. *Mutat. Res.* 635, 81-89.

Hinwood, A. L., Rodriguez, C., Runnion, T., Farrar, D., Murray, F., Horton, A., Glass, D., Sheppeard, V., Edwards, J. W., Denison, L., Whitworth, T., Eiser, C., Bulsara, M.,

Gillett, R. W., Powell, J., Lawson, S., Weeks, I., Galbally, I., (2007). Risk factors for increased BTEX exposure in four Australian cities. *Chemosphere* 66, 533-541.

Hunkeler, D., Höhener, P., Zeyer, J., (2002). Engineered and subsequent intrinsic *in situ* bioremediation of a diesel fuel contaminated aquifer. *J. Contam. Hydrol.* 59, 231-245.

Hutchins, S. R., Sewell, G. W., Kovacs, D. A., Smith, G. A., (1991). Biodegradation of aromatic hydrocarbons by aquifer microorganisms under denitrifying conditions. *Environ. Sci. Technol.* 25, 68-76.

IARC. International Agency for Research on Cancer, IARC Monograph, *Evaluation of Carcinogenic Risks to Humans, Some Industrial Chemicals*, vol. 77, 2000, p. 227-266.

Inazaki, T. H. *Avaliação da biodegradação de efluentes com substâncias persistentes antes e após tratamento eletrolítico para proposta de tratamento físico-biológico alternativo.* 127f. Dissertation, UNESP, 2003.

Irwin, R. J.; Vanmouwerik, M.; Stevens, L.; Seese, M. D.; Basham, W. *Environmental Contaminants Encyclopedia*, Fort Collins: Water Resources Division, 1997.

Jo, M. S., Rene, E. R., Kim, S. H., Park, H. S., (2008). An analysis of synergistic and antagonistic behavior during BTEX removal in batch system using response surface methodology. *J. Hazard. Mater.* 152, 1276-1284.

Johnson, E. M., Gabel, B. E. G., Christian, M. S., Sica, E., (1986). The developmental toxicity of xylene and xylene isomers in the *Hydra* assay. *Toxicol. Appl. Pharmacol.* 82, 323-328.

Johnson, S. J., Woolhouse, K. J., Prommera H., Barrya, D. A., Hristofi N., (2003). Contribution of anaerobic microbial activity to natural attenuation of benzene in groundwater. *Eng. Geol.* 2186, 1-7.

Kao, C. M., Wang, C. C., (2000). Control of BTEX migration by intrinsic bioremediation at a gasoline spill site. *Water Res.* 34, 3413-3423.

Kurás, M., Nowakowska, J., Sliwinska, E., Pilarski, R., Ilasz, R., Tykarska, T., Gulewicz, K., (2006). Changes in chromossome structure, mitotic activity and nuclear DNA content from cells of *Allium* test induced by barck water extract of *Uncaria tomentosa* (Willd.) DC. *Chemosphere* 107, 211-221.

Langman, J. M., (1994). Xylene: its toxicity, measurement of exposure levels, absorption, metabolism and clearance. *Pathology* 26, 301-309.

Lee, J. Y., Roh, J. R., Kim, H. S., (1994). Metabolic engineering of *Pseudomonas putida* for the simultaneous biodegradation of benzene, toluene, and p-xylene mixture. *Biotechnol. Bioeng.* 43, 1146-1152.

Lee, S. H., Jin, H. M., Lee, H. J., Kim, J. M., Jeon, C. O. (2012). Complete genome sequence of the BTEX-degrading bacterium *Pseudoxanthomonas spadix* BD-a59. *J. Bacteriol.* 194, 544.

Leite, E. A. Solventes orgânicos. In: OGA, S. (Ed.). *Fundamentos de toxicologia*. 1. ed. São Paulo: Atheneu, 1996, p.187-229.

Lin, C.-W., Chen, L.-H., Yet-Pole, I., Lai, C.-Y., (2010). Microbial communities and biodegradation in lab-scale BTEX-contaminated groundwater remediation using an oxygen-releasing reactive barrier. *Bioprocess Biosyst. Eng.* 33, 383-391.

Lovley, D. R., (1997). Potential for anaerobic bioremediation of BTEX in petroleum-contaminated aquifers. *J. Ind. Microbiol. Biot.* 18, 75-81.

Mackay, D.; Shiu, W. Y.; Ching Ma, K. *Illustrated Handbook of Physical-Chemical Properties and Environmental Fate for Organic Chemicals:* Volume I - Monoaromatic Hydrocarbons, Chlorobenzenes, and PCBs. Boca Raton: CRC Press, 697 p., 1992.

Maila, M. P., Cloete, T. E., (2005). The use of biological activities to monitor the removal of fuel contaminants – perspective for monitoring hydrocarbon contamination: a review. *Int. Biodeter. Biodegr.* 55, 1-8.

Mariano, A. P. *Avaliação do potencial de biorremediação de solos e águas subterrâneas contaminados com óleo diesel.* 147f. Thesis, UNESP, 2006.

Matsumoto, S. T., Mantovani, M. S., Malaguttii, M. I. A., Dias, A. L., Fonseca, I. C., Marin-Morales, M. A., (2006). Genotoxicity and mutagenicity of water contaminated with tannery effluents, as evaluated by the micronucleus test and comet assay using the fish *Oreochromis niloticus* and chromosome aberrations in onion root-tips. *Genet. Mol. Biol.* 29, 148-158.

Mazzeo, D. E. C., Levy, C. E., Angelis, D. F., Marin-Morales, M. A., (2010). BTEX biodegradation by bacteria from effluents of petroleum refinery. *Sci. Total Environ.* 408, 4334-4340.

Mazzeo, D. E. C., Fernandes, T. C. C., Marin-Morales, M. A., (2011). Cellular damages in the *Allium cepa* test system, caused by BTEX mixture prior and after biodegradation process. *Chemosphere* 85, 13-18.

Mazzeo, D. E. C., Matsumoto, S. T., Levy, C. E., Angelis, D. F., Marin-Morales, M. A., (2013). Application of micronucleus test and comet assay to evaluate BTEX biodegradation. *Chemosphere* 90, 1030-1036.

Mazzuco, L. M. *Atenuação natural de hidrocarbonetos aromáticos em aquíferos contaminados com óleo diesel.* 86f. 2004. Dissertation - UFSC, 2004.

Mcgregor, D., (1994). The genetic toxicology of toluene. *Mutat. Res.* 317, 213-228.

Mcgregor, D. B., Brown, A., Cattanach, P., (1988). Responses of the L5178Y tk+/tk- mouse lymphoma cell forward mutation assay: III. 72 Coded chemicals. *Environ. Mol. Mutagen.* 12, 85-154.

MEDITEXT. Medical Management. Xylene (mixed isomers). In: TOMES CPS tm SYSTEM. *Toxicology, occupational medicine and environmental series.* Englewood: Micromedex, 2000.

Melo, I. S.; Azevedo, J. L. *Microbiologia Ambiental.* Embrapa-CNPMA, 1997, 246p.

Migid, A. H. M., Azab, Y. A., Ibrahim, W. M., (2007). Use of plant genotoxicity bioassay for the evaluation of efficiency of algal biofilters in bioremediation of toxic industrial effluent. *Ecotoxicol. Environ. Safe.* 66, 57-64.

Miller, L., Xu, X., Grgicak-Mannion, A., Brook, J., A. Wheeler., (2012). Multi-season, multi-year concentrations and correlations amongst the BTEX group of VOCs in an urbanized industrial city. *Atmos. Environ.* 61, 305-315.

Mullin, A. H., Nataraj, D., Ren, J.-J, Mullin, D. A., (1998). Inhaled benzene increases the frequency and length of lacI deletion mutations in lung tissues of mice. *Carcinogenesis* 19, 1723-1733.

Murata, M., Tsujikawa, M., Kawanishi, S., (1999). Oxidative DNA damage by minor metabolites of toluene may lead to carcinogenesis and reproductive dysfunction. *Biochem. Bioph. Res. Commun.* 261, 478-483.

Nadim, F., Hoag, G. E., Liu, S., Carley, R. J., Zack, P., (2000). Detection and remediation of soil and aquifer systems contaminated with petroleum products: an overview. *J. Petrol. Sci. Eng.* 26, 169-178.

Nakhla, G., (2003). Biokinetic modeling of in situ bioremediation of BTX compounds – impact of process variable and scaleup implications. *Water Res.* 37, 1296-1307.

Nestmann, E. R., Lee, E. G. H., (1983). Mutagenicity of constituents of pulp and papermill effluent in growing cells of *Saccharomyces cerevisiae*. *Mutat. Res.* 119, 273-280.

Nikolova, N.; Nenov, V., (2005). BTEX degradation by fungi. *Water Sci. Technol.* 51, 87-93.

NLM. National Library of Medicine. *Hazardous Substances Data*, Denver: Micromedex Inc., 1988.

NLM. National Library of Medicine. *Hazardous Substances Data Bank*, Denver: Micromedex Inc., 1994.

NTP. National Toxicology Program. *Toxicology and carcinogenesis studies of xylenes (mixed) (60% m-xylene, 14% p-xylene, 9%-xylene, and 17% ethylbenzene) in F334/N rats and B6C3Fl mice (gavage studies).* Research Triangle Park: NTP. 1986.

NTP. National Toxicology Program. *Toxicology and carcinogenesis studies of ethylbenzene in F344/N rats and B6C3Fl mice.* Inhalation studies, Research Triangle Park: NTP, 1996, 466p.

Nunes-Halldorson, V. S., Steiner, R. L., Smith, G. B., (2004). Residual toxicity after biodegradation: interactions among benzene, toluene, and chloroform. *Ecotoxicol. Environ. Safe.* 57, 162-167.

OPS. Organización Panamericana de la Salud. Hidrocarburos aromáticos y salud: el estado del arte. In: MENDES, R. *Salud ocupacional en la industria del petróleo.* Bogotá, 1996.

OTA. Office of Technology Assessment. *Alternatives to animal use in research, testing and education.* Washington DC: US. Government Printing Office. 1986.

Patra, M., Sharma, A., (2002). Relative efficacy of *Allium cepa* and *Allium sativum* in anaphase-telophase test screening metal genotoxicity. *Biologia* 57, 409-414.

Pedrozo, M. F. M.; Barbosa, E. M.; Corseuil, H. X.; Schneider, M. R.; Linhares, M. M. *Ecotoxicologia e avaliação de risco do petróleo.* Salvador: NEAMA, 2002, 246p.

Phillips, T. M., Liu, D., Seech, A. G., Lee, H., Trevors, J. T., (2000). Monitoring bioremediation in creosote-contaminated soils using chemical analyses and toxicity tests. *J. Ind. Microbiol. Biot.* 24, 132-139.

Plaza, G., Nalecz-Jawecki, G., Ulfig, K., Brigmon, R. L., (2005). Assessment of genotoxic activity of petroleum hydrocarbon-bioremediated soil. *Ecotoxicol. Environ. Safe.* 62, 415-420.

Prenafeta-Boldú, F. X., Vervoort, J., Grotenhuis, J. T. C., Van Groenestijn, J. W., (2002). Substrate interactions during the biodegradation of benzene, toluene, ethylbenzene, and xylene (BTEX) hydrocarbons by the fungus *Cladophialophora* sp. strain T1. *Appl. Environ. Microbiol.* 68, 2660-2665.

Quinzani-Jordão, B. *Ciclo celular em meristemos. La formatión de intercâmbios entre cromátidas hermanas.* 1978. 276f. Thesis, Universidade de Complutense, Madrid.

Randerath, K., Randerath, E., Zhou, G. D., Suponpong, N., He, L. Y., Mcdonald, T. J., Donnelly, K. C., (1999). Genotoxicity of complex PAH mixtures recovered from contaminated lake sediments as assayed by three different methods. *Environ. Mol. Mutagen.* 33, 303-312.

Rank, J., Nielsen, M. H., (1994). Evaluation of the *Allium* anaphase–telophase test in relation to genotoxicity screening of industrial wastewater. *Mutat. Res.* 312, 17-24.

Rank, J., Nielsen, M. H., (1998). Genotoxicity testing of wastewater sludge using the *Allium cepa* anaphase-telophase chromosome aberration assay. *Mutat. Res.* 418, 113-119.

Régis, G. *Tratamento eletrolítico das águas residuárias de uma indústria produtora de antioxidante de borracha visando sua biodegradação*. 117f. Dissertation, UNESP, 2000.

Rene, E. R., Mohammad, B. T., Veiga, M. C., Kennes, C., (2012). Biodegradation of BTEX in a fungal biofilter: Influence of operational parameters, effect of shock-loads and substrate stratification. *Bioresource Technol.* 116, 204-213.

Reusser, D. E., Istok, J. D., Beller, H. R., Field, J. A., (2002). *In situ* transformation of deuterated toluene and xylene to benzylsuccinic acid analogues in BTEX-contaminated aquifers. *Environ. Sci. Technol.* 36, 4127-4134.

Ridgway, H. F., Safarik, J., Phipps, D., Carl, P., Clark, D., (1990). Identification and catabolic activity of well-derived gasoline-degrading bacteria from a contaminated aquifer. *Appl. Environ. Microbiol.* 56, 3565-3575.

Roberts, L. G., Nicolich, M. J., Schreiner, C. A., (2007). Developmental and reproductive toxicity evaluation of toluene vapor in the rat II. Developmental toxicity. *Reprod. Toxicol.* 23, 521-531.

Rogero, S. O., Lugão, A. B., Ikeda, T. I., Cruz, A. S., (2003). Teste *in vitro* de citotoxicidade: Estudo comparativo entre duas metodologias. *Mater. Res.* 6, 317-320.

Röling, W. F. M., Versevel, H. W., (2002). Natural attenuation: What does the subsurface have in store? *Biodegrad.* 13, 53-64.

Roma-Torres, J., Teixeira, J. P., Silva, S., Laffon, B., Cunha, L. M., Méndez, J., Mayan, O., (2006). Evaluation of genotoxicity in a group of workers from a petroleum refinery aromatics plant. *Mutat. Res.* 604, 19-27.

Safarova, V. I., Sapelnikova, S. V., Djazhenko, E. V., Teplova, G. I., Shajdulina, G. F., Kudasheva, F.K., (2004). Gas chromatography–mass spectrometry with headspace for the analysis of volatile organic compounds in waste water. *J. Chromatogr. B* 800, 325-330.

Saillenfait, A. M., Gallissot, F., Morel, G., Bonnet, P., (2003). Developmental toxicities of ethylbenzene, ortho-, meta-, para-xylene and technical xylene in rats following inhalation exposure. *Food Chem. Toxicol.* 41, 415-429.

Saillenfait, A. M., Gallissot, F., Sabaté, J. P., Bourges-Abella, N., Cadot, R., Morel, G., Lambert, A. M., (2006). Developmental toxicity of combined ethylbenzene and methylethylketone administered by inhalation to rats. *Food Chem. Toxicol.* 44, 1287-1298.

Shim, H., Hwang, B., Lee, S. S., Kong, S. H., (2005). Kinetics of BTEX biodegradation by a coculture of *Pseudomonas putida* and *Pseudomonas fluorescens* under hypoxic conditions. *Biodegrad.* 16, 319-327.

Sikkema, J., De Bont, J. A. M., Poolman, B., (1995). Mechanisms of membrane toxicity of hydrocarbons. *Microbiol. Rev.* 59, 201-222.

Silva, L. R., Ferreira, M. M., (2003). Estudo do coeficiente de partição octanol-água de bifenilas policloradas (PCBs) utilizando parâmetros topológicos. *Quím. Nova* 26, 312-318.

Silva, M. L. B., Corseuil, H. X., (2012). Groundwater microbial analysis to assess enhanced BTEX biodegradation by nitrate injection at a gasohol-contaminated site. *Int. Biodeter. Biodegr.* 67, 21-27.

Silva, R. L. B., Barra, C. M., Monteiro, T. C. N., Brilhante, O. M., (2002). Estudo da contaminação de poços rasos por combustíveis orgânicos e possíveis conseqüências para a saúde pública no Município de Itaguaí, Rio de Janeiro, Brasil. *Caderno Saúde Pública* 18, 1599-1607.

Smith, M. T., Rothman, N., (2000). Biomarkers in the molecular epidemiology of benzene-exposed workers. *J. Toxicol. Environ. Health* 61, 439-445.

Sousa, S., Duffy, C., Weitz, H., Glover, A. L., Bär, E., Henkler, R., Killham, K., (1998). Use of lux-modified bacterial biosensors to identify constraints to bioremediation of BTEX-contaminated sites. *Environ. Toxicol. Chem.* 17, 1039-1045.

Sul, D., Lee, D., Im, H., Oh, E., Kim, J., Lee, E., (2002). Single strand DNA breaks in T- and B-lymphocytes and granulocytes in workers exposed to benzene. *Toxicol. Lett.* 134, 87-95.

Tokunaga, I., Gotohda, T., Ishigama, A., Kitamura, O., Kubo, S., (2003). Toluene inhalation induced 8-hydroxy-2'-deoxyguanosine formation as the peroxidative degeneration in rat organs. *Legal Med.* 5, 34-41.

Tompa, A., Major, J., Jakab, M. G., (1994). Monitoring of benzene exposed workers for genotoxic effects of benzene: improvedworking-condition-related decrease in the frequencies of chromosomal aberrations in peripheral blood lymphocytes. *Mutat. Res.* 304, 159-165.

University of Washington. *Teratogen Information System*, Denver: Micromedex Inc., 1990.

USEPA. U.S. Environmental Protection Agency. *Oil and Hazardous Materials/Technical Assistance Data System (OHM/TADS).* Denver: Micromedex Inc.: 1985.

USEPA. U.S. Environmental Protection Agency. *Quality Criteria for Water.* Washington, D.C: Office of Water Regulations and Standards, 1986.

USEPA. U.S. Environmental Protection Agency. *Water Quality Criteria Summary List. Based on Quality Criteria for Water.* Denver: EPA Water Quality Office, 1991.

USEPA. U.S. Environmental Protection Agency. *Region IX preliminary remediation goals (PRGs) second half 1995.* San Francisco: USEPA, 1995a.

USEPA. U.S. Environmental Protection Agency. *Region III Risk- Based Concentration Table.* Philadelphia: USEPA, 1995b.

USEPA. U.S. Environmental Protection Agency. *Toxicological review of toluene (CAS No. 108-88-3).* Washington: USEPA, 2005, 107p.

Van't Hof, J., Schairer, L. A., (1982). *Tradescantia* assay system for gaseous mutagens: A report of the U.S. Environmental Protection Agency Gene-Tox Program. *Mutat. Res.* 99, 303-315.

Villatoro-Monzón, W. R., Mesta-Howard, A. M., Razo-Flores, E., (2003). Anaerobic biodegradation of BTEX using Mn(IV) and Fe(III) as alternative electron acceptors. *Water Sci. Technol.* 48, 125-131.

Walden, T., Spence, L., (1997). Risk-based BTEX screening criteria for a groundwater irrigation scenario. *Hum. Ecol. Risk Assess.* 3, 699-722.

Wang, X. M., Sheng, G. Y., Fu, J. M., Chan, C. Y., Lee, S. C., Chan, L. Y., Wang, Z. S., (2002). Urban roadside aromatic hydrocarbons in three cities of the Pearl River Delta, People's Republic of China. *Atmos. Environ.* 36, 5141-5148.

Waters, M. D., stack, H. F., Brady, A. L, Lohman, P. H. M., Haroun, L., Vainio, H., (1988). Use of computerized data listings and activity profiles of genetic and related effects in the review of 195 compounds. *Mutat. Res.* 205, 295-312.

WHO. World Health Organization. *Environmental Health Criteria 52: Toluene*, Geneva: WHO, 1986.

WHO. World Health Organization. *Environmental Health Criteria 150: Benzene*, Geneva: World Health Organization, 1993.

WHO. World Health Organization. *Environmental Health Criteria 190: Xylenes*, Geneva: WHO, 1997.

WHO. World Health Organization. *Air Quality Guidelines*, Copenhagen: WHO Regional Office for Europe, 2000.

WHO. World Health Organization. *Guidelines for drinking-water quality*, 4th Edition, Geneva: WHO, 2011.

Whysner, J., (2000). Benzene-induced genotoxicity, *J. Toxicol. Environ. Health* 61, 347-351.

Whysner, J., Vijayaraj Reddy, M., Ross, P. M., Mohan, M., Lax, E.A., (2004). Genotoxicity of benzene and its metabolites. *Mutat. Res.* 566, 99-130.

Xin, B.-P., Wu, C.-H., Wu, C.-H., Lin, C.-W., (2013). Bioaugmented remediation of high concentration BTEX-contaminated groundwater by permeable reactive barrier with immobilized bead. *J. Hazard. Mater.* 244-245, 765-772.

Yan, S., Zhou, Q., (2011). Toxic effects of *Hydrilla verticillata* exposed to toluene, ethylbenzene and xylene and safety assessment for protecting aquatic macrophytes. *Chemosphere* 85, 1088-1094.

Yardley-Jones, A., Anderson, D., Lovell, D. P., Jenkinson, P. C., (1990). Analysis of chromosomal aberrations in workers exposed to low level benzene. *Brit. J. Ind. Med.* 47, 48-51.

Yeh, C.-H., Lin, C.-W., Wu, C.-H., (2010). A permeable reactive barrier for the bioremediation of BTEX-contaminated groundwater: Microbial community distribution and removal efficiencies. *J. Hazard. Mater.* 178, 74-80.

Zhang, L., Mchale, C. M., Rothman, N., Li, G., Ji, Z., Vermeulen, R., Hubbard, A. E., Ren, X., Shen, M., Rappaport, S. M., North, M., Skibola, C. F., Yin, S., Vulpe, C., Chanock, S. J., Smith, M. T., Lan, Q., (2010). Systems biology of human benzene exposure. *Chem.-Biolog. Interact.* 184, 86-93.

Zhang, Y., Mu, Y., Liu, J., Mellouki, A., (2012). Levels, sources and health risks of carbonyls and BTEX in the ambient air of Beijing, China. *J. Environ. Sci.* 24, 124-130.

Zheng, S., Zhou, Q., Gao, J., Xiong, H., Chen, C., (2012). Behavioral alteration and DNA damage of freshwater snail *Bellamya aeruginosa* stressed by ethylbenzene and its tissue residue. *Ecotoxicol. Environ. Safe.* 81, 43-48.

Zobell, C. E., (1946). Action of microörganisms on hydrocarbons. *Bacteriol. Rev.* 10, 1-49.

In: Natural Gas and Petroleum
Editor: Constantin Cranganu

ISBN: 978-1-62808-535-8

Chapter 7

ANALYSIS OF CYCLIC VARIABILITY IN A NATURAL GAS SPARK IGNITION ENGINE

Asok K. Sen*

Richard G. Lugar Center for Renewable Energy and Department of Mathematical Sciences, Indiana University, Indianapolis, IN, US

ABSTRACT

This chapter presents a study of the cycle-to-cycle variations (CCV) in a natural gas spark ignition engine with and without exhaust gas recirculation (EGR). The presentation is divided into two parts. In the first part, no EGR is used, and the CCV of the indicated mean effective pressure (IMEP) are examined for different compression ratios (CR). The engine is operated at 1200 rpm and 1800 rpm, and four values of CR = 8, 10, 12, and 14 are used. In the second part, the EGR level is changed from 0% to 5%, 10%, 15% and 20%, at a fixed compression ratio of 9.4. The engine is operated at 2000 rpm and a stoichiometric fuel-air mixture is used. In both parts, the IMEP time series are analyzed using a continuous wavelet transform. The dominant oscillatory modes of the CCV are identified and the engine cycles over which these modes may persist are delineated. The results reveal that the CCV of the IMEP occur on multiple timescales and exhibit complex dynamics. With no EGR, high frequency intermittent fluctuations are observed at higher speeds. As EGR level is increased, persistent low frequency variations tend to develop. In addition, the spectral power increased with an increase in the EGR level. At the EGR level of 20%, the spectral power is found to increase significantly indicating that EGR has a pronounced effect on increasing the CCV. Knowledge of the dominant modes of variability may be useful to develop effective control strategies for reducing the CCV and improving engine performance in the presence and absence of EGR.

Keywords: Natural gas, Cycle-to-cycle variations, Spark ignition engine, Exhaust gas recirculation, Wavelet analysis

---

* Corresponding author: Asok K. Sen. 402 N. Blackford Street, Indianapolis, IN 46202, US. E-mail: asen@iupui.edu.

INTRODUCTION

Currently there is a great deal of interest in developing efficient strategies to reduce exhaust emissions and improve thermal efficiency of internal combustion engines. To this end, alternative fuels and fuel blends are increasingly being used, replacing conventional fuels such as gasoline and diesel (Poulton, 1999; Kowalewicz and Wojtyniak, 2005; Gravalos et al., 2011). Among the various alternative fuels, natural gas has been used extensively in recent years (Cho and He, 2007). The main constituent of natural gas is methane, which has the highest hydrogen to carbon ratio compared to other hydrocarbon fuels, making natural gas one of the cleanest fuels available today. Natural gas has been used as a fuel in both spark-ignition engines and compression-ignition engines. In comparison to a gasoline engine, a natural gas engine can reduce NOx, CO, CO_2, and total hydrocarbon emissions to 35-60%, 90-97%, 25% and 50-75%, respectively (Evans and Blaszczyk, 1997; Shariat, 2007). Compared to a diesel engine, a natural gas engine can lead to a reduction of NOx, total hydrocarbon, and particulate matter emissions to 98%, 67% and 96%, respectively (Environmental Protection Agency, 2002). In addition, by virtue of its good antiknock properties, natural gas allows higher compression ratios to be used leading to higher thermal efficiency.

To comply with the more stringent emission requirements, the amount of exhaust gases, particularly NOx, produced by a natural gas spark ignition engine needs to be reduced further. The exhaust gas circulation (EGR) is one of the techniques that have been used effectively for this purpose (Tabata et al., 1995; Sakonji and Saoji, 1997; Alla, 2002; Qi et al., 2007; Saanum et al., 2007; Dimopoulos, 2007, 2008; Ibrahim and Bari, 2008; Hu et al., 2009a,b; Yoshioka et al., 2010). NOx is formed in high concentrations when the combustion temperature exceeds about 2500^0F. The EGR system works by recycling part of the exhaust gases into the engine cylinders. In a spark ignition engine, the recirculated inert exhaust gases displace some of the intake charge, thereby reducing the combustion temperature and NOx formation. In a compression ignition engine, the exhaust gases replace some of the excess oxygen in the pre-combustion mixture. The flow rate of the recirculating exhaust gases is controlled by a valve. It is important to regulate this flow rate. Because EGR dilutes the combustible mixture, a high EGR level may cause a misfire and unstable engine operation. In a typical spark ignition engine, about 5% to 20% of the exhaust gases are fed back into the intake manifold.

It has long been recognized that in an internal combustion engine the process variables such as pressure undergo cycle-to-cycle variations (CCV) (Young, 1981; Heywood, 1988; Daw et al., 1996; Ishii et al., 1997; Han, 2000; Lu et al., 2007; Daw et al., 2008). These cyclic variations may become severe under lean burn conditions or for highly dilute mixtures with exhaust gas recirculation (EGR) (Heywood, 1988). The CCV may lead to reduced power output, operational instabilities, and higher emissions. The CCV have been observed in spark ignition, compression ignition and homogeneous charge compression ignition (HCCI) engines. It has been estimated that in a spark ignition engine, elimination of the CCV may result in a 10% increase in power output for the same fuel consumption (Ozdor et al., 1994). It is important to understand the nature and dynamics of the CCV in order to develop effective strategies for controlling them (Kaul et al., 2009; Vance et al., 2009).

Due to the slow-burning velocity of natural gas, the natural gas spark ignition engine undergoes relatively large cycle-by-cycle variations in the lean mixture combustion and has a poor lean-burn capability. These restrictions may limit the operating range of the engine and increase its fuel consumption (Das and Watson, 1997; Akansu et al., 2004).

Several investigators have examined the CCV in natural gas spark ignition engines without EGR (Zervas, 2004; Huang et al., 2008; Wang et al., 2008; Ma et al., 2008; Sen et al., 2009; Zheng et al., 2009a,b). Zheng et al. (2009a,b) conducted experimental studies of CCV in a natural gas direct-injection spark-ignition engine with no EGR. Among other things, they analyzed the effect of compression ratio on the CCV of the indicated mean effective pressure (IMEP), and explored the interdependency between IMEP and flame development duration, main combustion duration, and total combustion duration. They used the coefficient of variation (COV) of the IMEP time series to study the effect of compression ratio on the CCV. Their analysis revealed that the CCV in a natural gas direct-injection spark-ignition engine may result from the cyclic variations in flame development duration, main combustion duration and total combustion duration.

The cycle-to-cycle variations (CCV) in a natural gas engine with EGR have also been examined by a few researchers (see Chang, 2002; Huang et al, 2009). Huang et al. (2009) used a spark ignition engine fueled by a stoichiometric mixture, and analyzed the IMEP time series for different rates of EGR. Use of a stoichiometric mixture is appropriate for such studies because with many EGR systems, a three-way catalyst is used for after-treatment of the exhaust gases, and this three-way catalyst works most efficiently with a stoichiometric mixture. Using the COV of the IMEP time series, they found that as the EGR level is increased, the CCV of the IMEP time series also increase.

The coefficient of variation (COV) of a time series is the ratio of its standard deviation to its mean value, which is a useful statistic for comparing the degree of variation between two time series even when their mean values are quite different from each other. For a given time series, the COV provides a single numerical measure characterizing the temporal variability in the data; however, it cannot yield any information about the spectral characteristics of the time series. The objective of this chapter is to explore the spectral and temporal variability of the IMEP time series in a natural gas engine with and without EGR, using wavelet analysis. Wavelet-based techniques are increasingly being used for time series analysis in a variety of applications (Addison, 2002). They are particularly useful for the analysis of transient and intermittent processes. In our recent work, we used a continuous wavelet transform (CWT) to analyze the cyclic variability in internal combustion engines (Sen et al., 2008, 2009, 2010, 2011a,b,c). With CWT, the dominant modes of variability can be identified and the engine cycles over which these modes may persist can be delineated. In this chapter, we follow the development of (Sen et al., 2011b,c) to describe the cyclic variability in a natural gas spark ignition engine in the presence and absence of EGR.

EXPERIMENTAL PROCEDURE AND ANALYSIS METHODOLOGY

Experimental Procedure

For the analysis of cyclic variability in a natural gas spark ignition engine with no EGR, we will use the IMEP time series that were obtained experimentally by Zheng et al. (2009b). These investigators used a single-cylinder engine which was run at 1200 and 1800 rpm.

A spark plug was installed inside the combustion chamber, and a production-type gasoline swirl injector was used. The flow rate of the injector was maintained at 191.8 liters/ min under an injection pressure of 8 MPa.

The natural gas used in the experiment consisted of 96.16% of methane by volume. The compression ratio was varied by modifying the piston crown. Four compression ratios CR = 8, 10, 12, and 14 were used. The injection timing, spark timing, and injection duration were controlled by an electronic control system, and the experiments were carried out at fixed fuel injection timing and fixed ignition timing. The engine was operated with 70% wide open throttle (WOT).

An eddy current dynamometer was used to measure the engine torque and speed. The in-cylinder pressure was measured by means of a Kistler 6117BF17 piezoelectric pressure transducer, and was recorded for every 0.1 degree of crank angle. The crank angle signal was acquired with a Kistler 2613B crank angle encoder, and the pressure and crank angle information were collected by a Yokogawa data acquisition system.

Further details about the experimental can be found in (Zheng et al., 2009b). The pressure measurements were made for 100 consecutive engine cycles. From the pressure measurements, the IMEP, which is the average pressure in the cylinder over an engine cycle, is obtained from the formula:

$$IMEP = \frac{W_c}{V_d}. \tag{1}$$

here V_d is the engine displacement volume, and W_c is the amount of work done per cycle, defined as:

$$W_c = \oint P dV, \tag{2}$$

P being the actual (i.e., measured) pressure in the engine cylinder.

To investigate the effect of EGR on the CCV of the IMEP, we will use the experimental data of Huang et al. (2009). In their experiment, the engine was operated with a stoichiometric mixture at 2000 rpm, and a compression ratio of 9.4 was used. The experiments were conducted with optimum ignition timing which was regulated to the minimum ignition advance at which the engine has the maximum brake torque (MBT). A cooled EGR system was used.

After passing through the EGR cooler and the EGR valve, the exhaust gas entered the intake manifold and mixed with the fresh fuel-air mixture, Different EGR ratios could be set by adjusting the EGR valve. An ECM EGR5230 analyzer was used to monitor the EGR ratio. The EGR ratio is defined by the formula: $\%EGR = [V_{EGR} / (V_1 + V_{EGR})] \times 100$, where V_{EGR} is the volume flow rate of EGR, and V_1 is the volume flow rate of the intake fuel-air mixture. The EGR level was changed from 0% to 5%, 10%, 15% and 20%, and the pressure measurements were made over 200 consecutive engine cycles (see Huang et al., 2009 for further details).

Analysis Methodology

As mentioned in the introduction, we investigated the CCV of the IMEP time series with and without EGR using a continuous wavelet transform (CWT). The CWT maps the time series on to a time scale (or time-period) plane called the wavelet power spectrum (WPS) from which the dominant modes of variability and the engine cycles over which these dominant modes may persist can be discerned by visual inspection. The WPS is obtained as a convolution of the time series with a mother wavelet, which contains a scale parameter. The scale parameter controls the dilation and contraction of the mother wavelet. The smaller scales track the high-frequency features of the time series, whereas the low-frequency variations are followed by the larger scales. The wavelet transform is computed by manipulating the mother wavelet over the time series in two ways: it is moved to various locations, and it is stretched or squeezed. If the wavelet matches the shape of the time series at a specific scale and location, then a large transform value is obtained. If, however, the wavelet and the time series do not correlate well, a low value of the transform will result. The amount of signal energy contained at a specific scale s and location n is defined as the wavelet power spectrum (WPS), and is given by the squared modulus of the CWT. For time series with periodic and quasiperiodic components, Morlet wavelet is often used as the mother wavelet. The Morlet wavelet consists of a plane wave modulated by a Gaussian function. It contains a parameter which is usually referred to as the center frequency. This parameter determines the frequency and time resolutions of the corresponding CWT. A larger value of this parameter provides a higher frequency resolution, and a smaller value leads to a higher time resolution. In our computations we have used a Morlet wavelet with center frequency 6 as the mother wavelet. This choice provides a good balance between time and frequency localizations. For this choice, the scale is also approximately equal to the Fourier period and thus the terms scale and period can be used interchangeably for interpreting the results.

Additional information about the spectral properties of the IMEP time series can be obtained from the global wavelet spectrum (GWS), which is calculated by averaging the WPS at each scale over all engine cycles. The GWS displays power as a function of period or frequency and is analogous to a smoothed Fourier spectrum of the time series. The dominant spectral modes of the IMEP time series can be estimated from the various peaks in the GWS.

RESULTS AND DISCUSSION

Natural Gas Spark Ignition Engine with No EGR

First we consider the natural gas spark ignition engine with no EGR. Figure 1 shows the IMEP time series for the compression ratio: CR = 8, 10, 12, and 14, with the engine running at 1200 rpm. The wavelet power spectra (WPS) and the global wavelet spectra (GWS) of these time series are depicted in Figure 2. A colorbar representing the different power levels is shown below the WPS for CR= 8. The area below the U-shape curve denotes the cone of influence (COI). Because the time series are zero-padded at both ends for the computation of the WPS, power is reduced at both ends of the WPS.

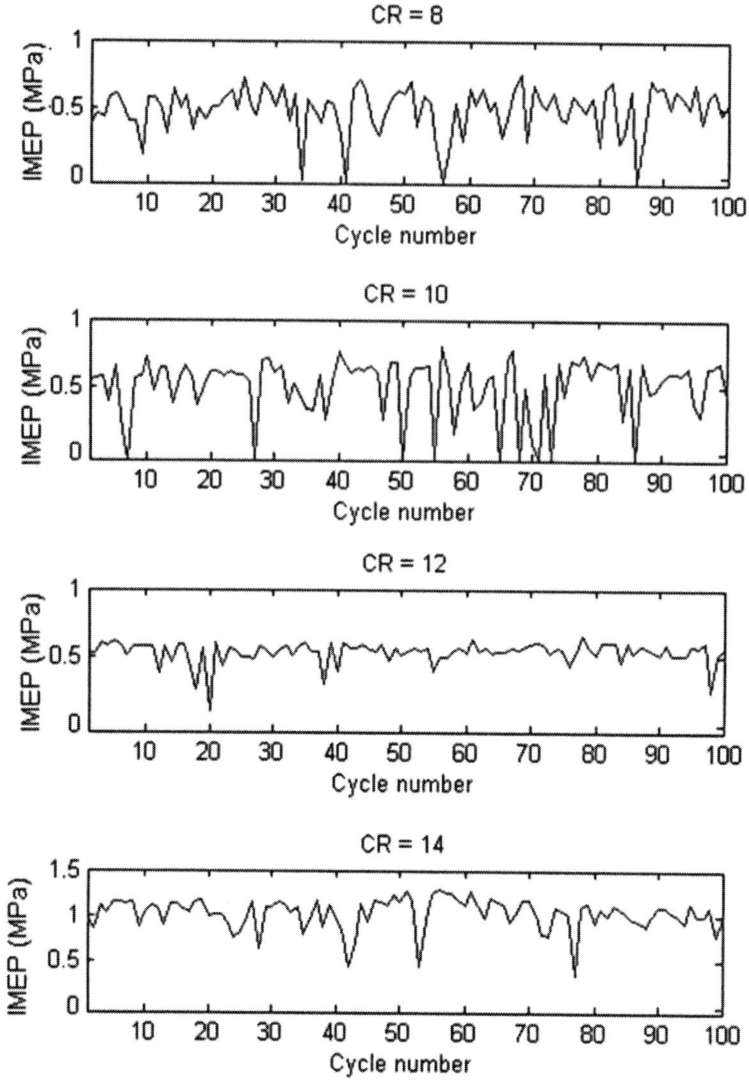

Figure 1. IMEP time series of the natural gas spark ignition engine with no EGR, for compression ratios CR = 8, 10, 12, and 14, at engine speed of 1200 rpm.

As a consequence, the results inside the COI may be unreliable and should be used with caution (see Torrence and Compo, 1998 for details). Due to limited record length of 100 cycles, we confine our considerations to periodicities less than 32-cycle.

Several characteristic features of the cyclic IMEP variations can be observed from Figure 2. The cyclic variations exhibit multiscale behavior with the fluctuations occurring at different timescales. At each compression ratio, there is significant power in various periodic bands.

The number of engine cycles over which these periodic bands persist is also evident from the WPS. We see that for CR = 8, there is significant power in the 8-12 cycle band, and also in the 16-24 cycle band, and they persist over 31-47 cycles and 27-87 cycles, respectively.

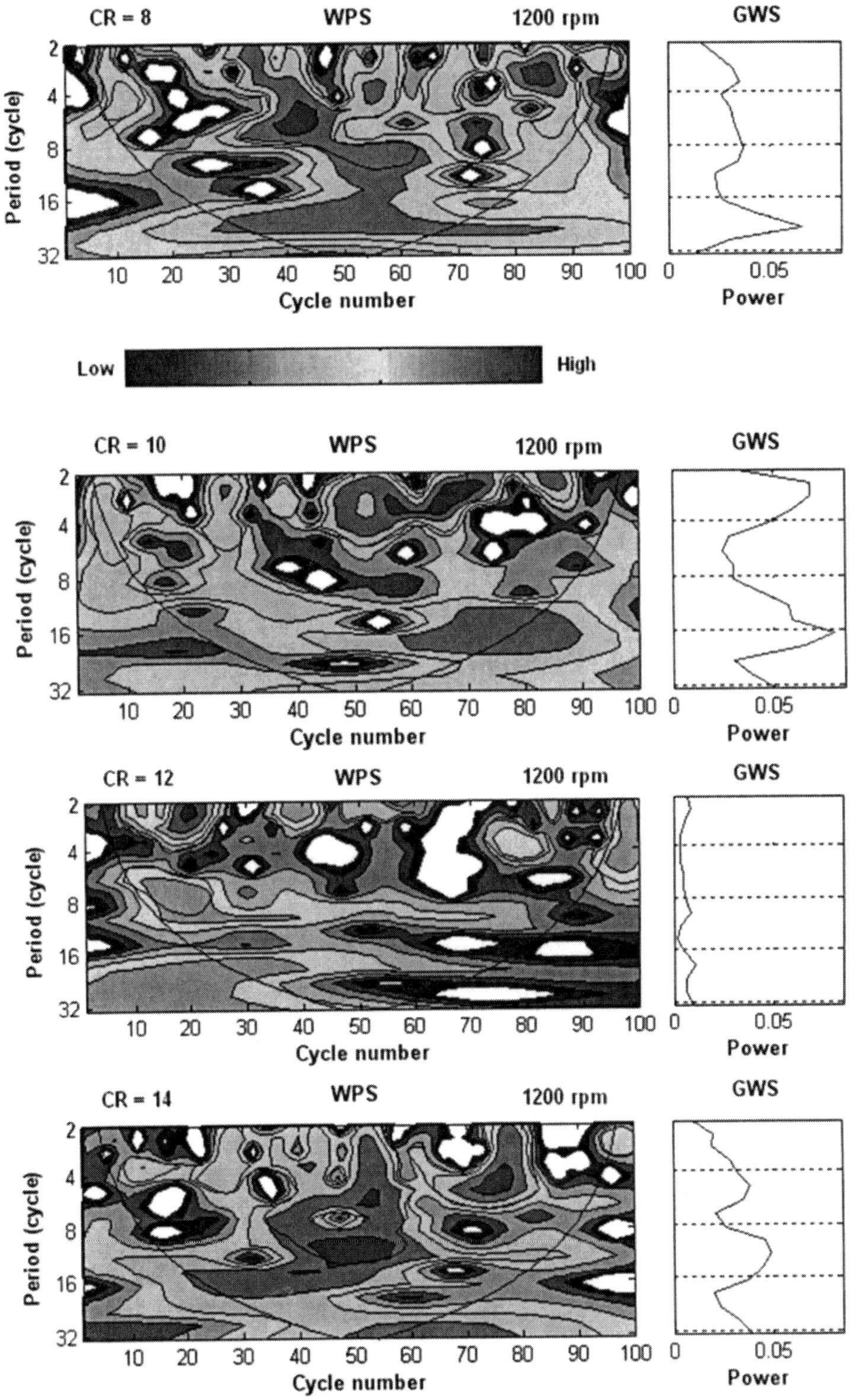

Figure 2. Wavelet power spectrum (WPS) and global wavelet spectrum (GWS) of the IMEP time series depicted in Figure 1.

Table 1. The periodic bands and the engine cycles over which these periodic bands persist. Engine speed = 1200 rpm

Compression Ratio (CR)	Periodic Band	No. of engine Cycles
CR = 8	2-4 cycle	78-88 cycles
	4-8cycle	31-47 cycles
	8-12 cycle	47-65 cycles
	16-24 cycle	27-87 cycles
CR = 10	2-4 cycle	48-76 cycles
	8-14 cycle	24-48 cycles
	12-20 cycle	40-80 cycles
CR = 12	2-3 cycle	16-24 cycles
	~10 cycle	16-38 cycles
CR = 14	3-6 cycle	66-82 cycle
	5-6 cycle	38-58 cycle
	8-12 cycle	38-60 cycle
	12-18 cycle	20-56 cycle

As the compression ratio is increased to 10 and 12, the low frequency bands weaken in their strength or tend to disappear, and when CR = 14, significant power reappears in the low frequency bands such as the 8-12 cycle, and 12-18 cycle, in a more continuous fashion. The periodic bands and the engine cycles over which they persist are listed in Table 1. The periodicity corresponding to the peak power in each periodic band may be discerned from the GWS shown on the right.

A comparison of the GWS for the different values of CR shown in Figure 2 reveals that the peak spectral power of cyclic IMEP variations is significantly reduced at CR = 12. This is consistent with the conclusion reached by Zheng et al. (2009b) on the basis of the coefficient of variation (COV) of the IMEP time series.

As the compression ratio increases, the cylinder gas pressure and temperature at the end of the compression stroke, the injected fuel concentration distribution in the cylinder and turbulence intensity of squish will be improved.

In addition, the amount of residual gases in the cylinder will decrease with increasing compression ratio. This will lead to an increase of the burning velocity of the mixture and thus a decrease of cycly-to-cycle variations. However, when the compression ratio is too large, the distribution of the injected fuel concentration in the cylinder will be unfavorable to mixture preparation and combustion, and this will lead to an increase in cycly-to-cycle variations. Therefore, the CCV of IMEP reaches the minimum value at an intermediate value of CR = 12.

For comparison, we present the corresponding results at the engine speed of 1800 rpm. The IMEP time series at this speed are shown in figure 3 for the four compression ratios. The WPS and GWS of these time series are illustrated in Figure 4.

From this figure, we see that as the speed is increased from 1200 rpm to 1800 rpm, strong high frequency oscillations appear and they occur intermittently along with more persistent low frequency variations. The various periodicities of these time series and the number of engine cycles over which they occur are listed in table II.

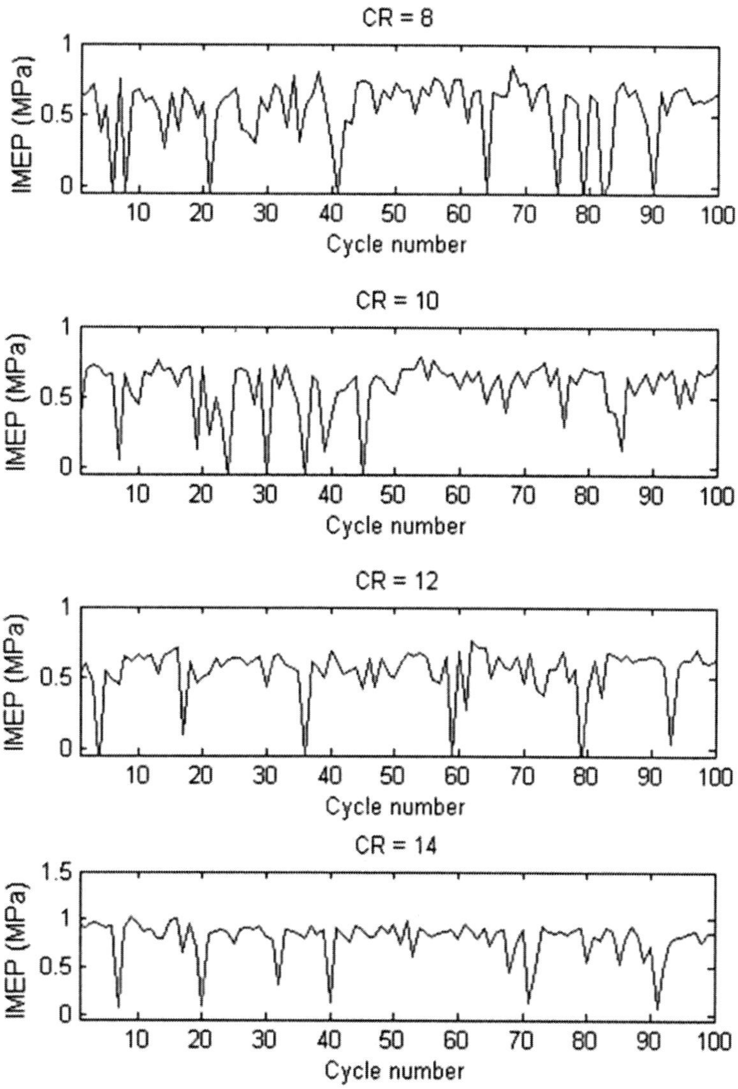

Figure 3. IMEP time series of the natural gas spark ignition engine with no EGR, for compression ratios CR = 8, 10, 12, and 14, at engine speed of 1800 rpm.

Natural Gas Spark Ignition Engine with EGR

Next we discuss the results for the natural gas spark ignition engine with EGR. Figure 5 shows the IMEP time series of this engine with 0%, 10%, and 20% EGR, at the engine speed of 2000 rpm. The WPS and GWS of each of these time series are depicted in Figure 6. The top panel in this figure shows that, in the absence of EGR, the CCV exhibit high frequency intermittent fluctuations. As EGR is introduced, persistent low frequency oscillations tend to develop in the CCV. For example, with EGR = 10%, there is strong power around the 16-cycle period which persists approximately over 60 engine cycles (see the third panel in Figure 6).

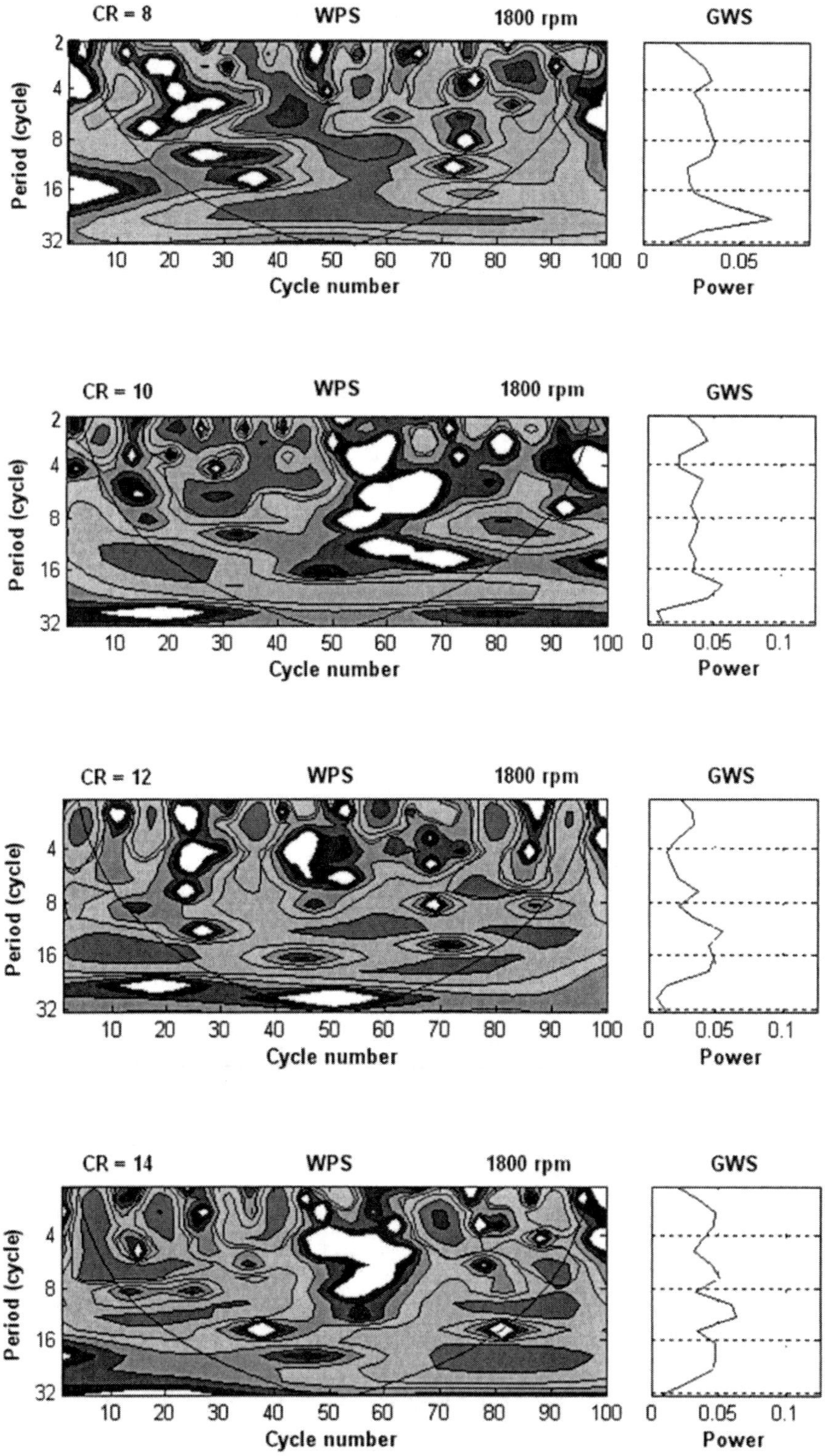

Figure 4. Wavelet power spectrum (WPS) and global wavelet spectrum (GWS) of the IMEP time series depicted in Figure 3.

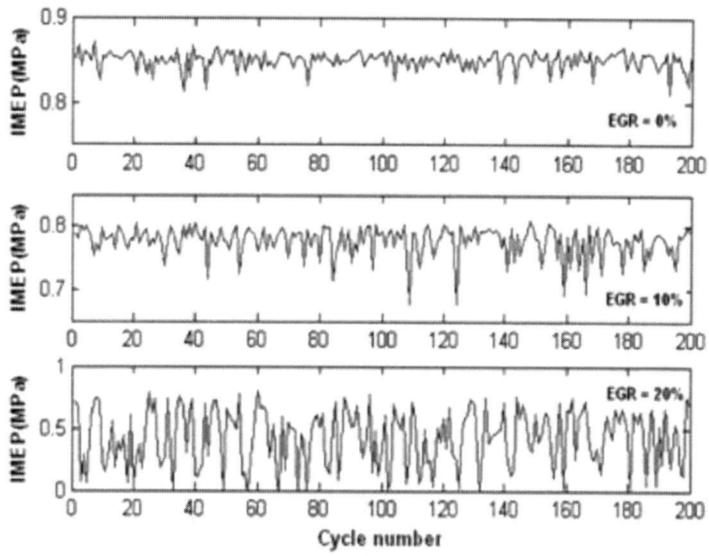

Figure 5. IMEP time series of the natural gas spark engine with EGR levels of 0%, 10%, and 20%, at engine speed of 2000 rpm.

Table 2. The periodic bands and the engine cycles over which these periodic bands persist. Engine speed = 1800 rpm

Compression Ratio (CR)	Periodic Band	No. of engine cycles
CR = 8	2-3 cycle	intermittent
	3-5 cycle	72-86 cycles
	6-8 cycle	10-45 cycles
CR = 10	2-3 cycle	18-46 cycles
	4-8 cycle	24-48 cycles
CR = 12	2-4 cycle	intermittent
	10-13 cycle	42-67 cycle
CR = 14	2-4 cycle	intermittent
	~ 7 cycle	4-25 cycles
	10-13 cycles	73-95 cycles

In addition, as seen from the GWS plots for different EGR levels, the spectral power increases as the EGR level increases. When the EGR level reaches 20%, we see from the bottom panel in Figure 6 that there is persistent oscillation around the 8-cycle period lasting over almost 60 engine cycles.

In addition, we find from the GWS in this figure that with EGR = 20%, the spectral power has increased significantly indicating that EGR has a pronounced effect on increasing the CCV of the IMEP.

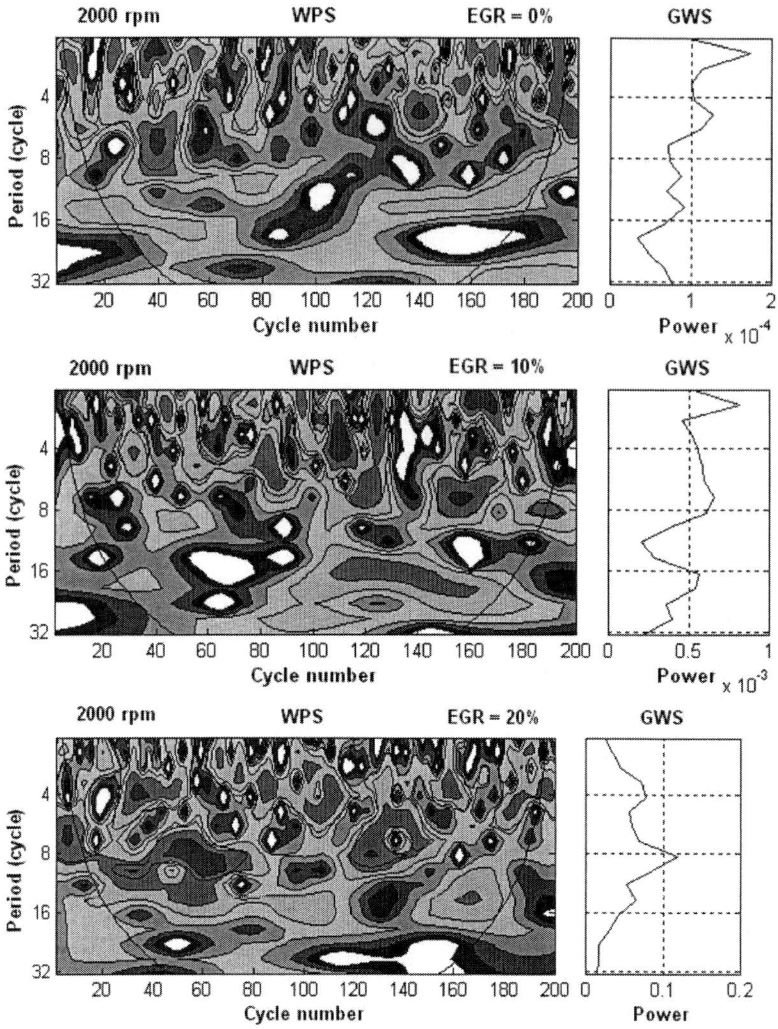

Figure 6. Wavelet power spectrum (WPS) and global wavelet spectrum (GWS) of the IMEP time series depicted in Figure 5.

CONCLUSION

We have examined the cycle-to-cycle variations (CCV) of the IMEP in a natural gas spark ignition engine with and without EGR. Using a continuous wavelet transform (CWT) we have identified the dominant periodicities in the IMEP time series and delineated the engine cycles over which these periodicities may persist. In the absence of EGR, our results indicate that at low speeds, the CCV exhibit persistent low frequency oscillations which prevail over several engine cycles; at higher speeds, high frequency intermittent fluctuations appear. When EGR is applied, persistent low frequency variations also tend to develop. In addition, as the EGR level is increased, the overall spectral power increases. At the EGR level of 20%, the spectral power is found to increase significantly indicating that EGR has a pronounced effect on increasing the CCV.

Knowledge of the dominant modes of variability may be useful to develop effective control strategies for reducing the CCV and improving engine performance. There have been several studies on reducing the CCV in the presence of EGR. To this end, use of natural gas-hydrogen blends as fuels has proven to be very effective (Hu et al, 2009a,b; Wang et al., 2009; Allenby et al., 2011).

Due to limited record length of the IMEP time series, we confined our considerations to periodicities less than 32-cycle. By conducting experiments over a larger number of engine cycles and analyzing the data, it would be possible to detect the presence of lower frequencies in the CCV. A longer record length will also enable us to determine if the periodicities identified here will persist over a longer time interval, i.e., more engine cycles.

ACKNOWLEDGMENT

I would like to thank Dr. Zuohua. Huang for providing the experimental data and previous collaborations.

REFERENCES

Addison, P. S., (2002). *The Illustrated Wavelet Transform Handbook*. Institute of Physics Publishing, Philadelphia.

Akansu, S. O., Dulger, A., Kahraman, N., (2004). Internal combustion engines fueled by natural gas-hydrogen mixtures. *Int. J. Hydrogen Energy* 29, 1527-1539.

Alla, G. H. A., (2002). Using EGR in an internal combustion engine: A review. *Energy Convers. Manage.* 43, 1027-1043.

Allenby, S., Chang, W. C., Megaritis, A., Wyszynski, A. J., (2011). Hydrogen enrichment: A way to maintain combustion stability in a natural gas fueled engine with exhaust gas recirculation. *Proc. Inst. Mech. Engrs Part D: J. Auto. Eng.* 215, 405-418.

Chang, W. C., (2002). An improved method of investigation of combustion parameters in a natural gas fueled SI engine with EGR and H_2 as additive. *PhD Thesis*, University of Birmingham, United Kingdom.

Cho, H. M., He, B. Q., (2007). Spark ignition natural gas engines – A review. *Energy Convers. Manage.* 48, 608-618.

Das, A., Watson, H. C., (1997). Development of a natural gas spark ignition engine for optimum performance. *Proc. Inst. Mech. Engrs Part D: J. Auto. Eng.* 211, 361-378.

Daw, C. S., Edwards, K. D., Wagner, R. M., Green, J. B., (2008). Modeling cyclic variability in spark-ignited HCCI. *J. Eng. Gas Turb. Power* 130:052801.

Daw, C. S., Finney, C. E. A., Green, J. B., Kennel, M. B., Thomas, J. F., Connolly, F. T., (1996). A simple model for cyclic variations in a spark-ignition engine. *SAE Paper No. 962086.*

Dimopoulos, P., Rechsteiner, C., Soltic, P., Laemmle, C., Boulouchos, K., (2007). Increase of passenger car engine efficiency with low engine-out emissions using hydrogen–natural gas mixtures: a thermodynamic analysis. *Int. J. Hydrogen Energy* 32, 3073–3083.

Dimopoulos, P., Bach, C., Soltic, P., Boulouchos, K., (2008). Hydrogen–natural gas blends fuelling passenger car engines: combustion, emissions and well-to-wheels assessment. *Int. J. Hydrogen Energy* 33, 7224–7236.

Environmental Protection Agency (EPA), (2002). US Clean alternative fuels: compressed natural gas: fact sheet, http://www.epa.gov/oms/consumer/fuels/altfuels/ 420f00033.pdf.

Evans, R. L., Blaszczyk, J., (1997). A comparative study of the performance and exhaust emissions of a spark ignition engine fueled by natural gas and gasoline. *Proc. Inst. Mech. Engrs. Part D: J. Auto. Eng.* 211, 39-47.

Gravalos, I., Moshou, D., Gialamas, T., Xyradakis, P., Kateris, D., Tsiropoulos, Z., (2011). Performance and emission characteristics of spark ignition engine fuelled with ethanol and methanol gasoline blended fuels, In: *Alternative Fuels*, Chapter 7, M. Manzanera (editor), InTech Publications, New York.

Han, S. B., (2000). Cycle-to-cycle variations under cylinder pressure-based combustion analysis in spark ignition engines. *J. Mech. Sci. Tech.* 14, 1151-1158.

Heywood, J. B., (1988). *Internal Combustion Engine Fundamentals*. McGraw Hill, New York.

Hu, E., Huang, Z., Liu, B., Zheng, J., Gu, X., Huang, B., (2009a). Experimental investigation on performance and emissions of a spark-ignition engine fueled with natural gas–hydrogen blends combined with EGR. *Int. J. Hydrogen Energy* 34, 528–539.

Hu, E., Huang, Z., Liu, B., Zheng, J., Gu, X., (2009b). Experimental study on combustion characteristics of a spark ignition engine fueled with natural gas–hydrogen blends combining with EGR. *Int. J. Hydrogen Energy* 34, 1035–1044.

Huang, Z., Liu, L., Jiang, D. M., Ren, Y., Liu, B., Zeng, K., Wang, Q., (2008). Study on cycle-by-cycle variations of combustion in a natural-gas direct-injection engine. *Proc. Inst. Mech. Engrs Part D: J. Auto. Eng.* 222, 1657-1667.

Huang, B., Hu, E., Huang, Z., Zheng, J., Liu, B., Jiang, D. M., (2009). Cycle-by-cycle variations in a spark ignition engine fueled with natural gas-hydrogen blends combined with EGR. *Int. J. Hydrogen Energy* 34, 8405-8414.

Ibrahim, A., Bari, S., (2008). Optimization of a natural gas spark ignition engine employing EGR strategy using a two-zone combustion model. *Fuel* 87, 1824-1834.

Ishii, K., Sasaki, T., Urata, Y., Yoshida, K., Ohno, T., (1997). Investigation of cyclic variation of IMEP under lean burn operation in spark ignition engines. *SAE Paper No. 972830.*

Kaul, B. C., Vance, J. B., Drallmeier, J. A., Sarangapani, J., (2009). A method for predicting performance improvements with effective cycle-to-cycle control of highly dilute spark ignition engine combustion. *Proc. Inst. Mech. Engrs Part D: J. Auto. Eng.* 223, 423-438.

Kowalewicz, A., Wojtyniak, M., (2005). Alternative fuels and their application to combustion engines. *Proc. Inst. Mech. Engrs Part D: J. Auto. Eng.* 219, 103-125.

Lu, X., Ji, L., Ma, J., Huang, Z., (2007). Experimental study on the cycle-by-cycle variations of homogeneous charge compression ignition combustion using primary reference fuels and their mixtures. *Proc. Inst. Mech. Engrs Part D: J. Auto. Eng.* 221, 859-866.

Ma, F., Wang, Y., Liu, H., Li, Y., Wang, J., Ding, S., (2008). Effects of hydrogen addition on cycle-by-cycle variations in a lean burn natural gas spark-ignition engine. *Int. J. Hydrogen Energy* 33, 823–831.

Ozdor, N., Dulger, M., Sher, E., (1994). Cyclic variability in spark ignition engines: A literature survey. *SAE Paper No. 940987.*

Poulton, M. L., (1999). *Alternative Fuels for Road Vehicles*. Computational Mechanics Publications, Bellerica, Massachusetts.

Qi, Y., Srinivasan, K. K., Krishnan, S. R., Yang, H., Midkiff, K. C., (2007). Effect of hot EGR on the performance and emissions of an advanced injection low pilot-ignited natural gas engine. *Int. J. Engine Res.* 8, 289-305.

Saanum, I., Tunestat, P., Johansson, B., (2007). Lean burn versus stoichiometric operation with EGR and 3-way catalyst of an engine fueled with natural gas and hydrogen enriched natural gas. *SAE Paper No. 2007-01-0015.*

Sakonji, T., Shoji, F., (1997). Study on improvement of natural gas fueled spark ignition engines – effect of EGR utilization. *SAE Paper No. 971714.*

Sen, A. K., Longwic, R., Litak, G., Gorski, K., (2008). Analysis of cycle-to-cycle pressure oscillations in a diesel engine. *Mech. Syst. Signal Process.* 22, 362-373.

Sen, A. K., Litak, G., Yao, B. F., Li, G. X., (2009). Analysis of pressure fluctuations in a natural gas engine under lean burn conditions. *Appl. Therm. Eng.* 30, 776-779.

Sen, A. K., Litak, G., Finney, C. E. A., Daw, C. S., Wagner, R. M., (2010) Analysis of heat release dynamics in an internal combustion engine using multifractals and wavelets. *Appl. Energy* 87, 1736-1743.

Sen, A. K., Litak, G., Edwards, K. D., Finney, C. E. A., Daw, C. S., Wagner, R. M., (2011a). Characteristics of cyclic heat release variability in the transition from spark ignition to HCCI in a gasoline engine. *Appl. Energy* 88, 1648-1655.

Sen, A. K., Zheng, J., Huang, Z., (2011b). Dynamics of cycle-to-cycle variations in a natural gas direct-injection spark-ignition engine. *Appl. Energy* 88, 2324- 2334.

Sen, A. K., Ash, S. K., Huang, B., Huang, Z., (2011c). Effect of exhaust gas recirculation on cyclic variability in a natural gas spark-Ignition engine. *Appl. Therm. Eng.* 31, 2247-2253.

Shariat, B. S., (2007). A comparative evaluation of SI engines fueled by natural gas and gasoline. *Second ANGVA Conf.*, November 27-29, Bangkok, Thailand.

Tabata, M., Yamamoto, T., Fukube, T., (1995). Improving NOx and fuel economy for a mixture-injected SI engine with EGR. *SAE Paper No.* 950684.

Torrence, C., Compo, G. P., (1998). A practical guide to wavelet analysis. *Bull. Amer. Meteorol. Soc.* 79, 61-78.

Vance, J., Kaul, B., Jagannathan, S., Drallmeier, J., (2009). Neuro emission controller for minimizing cyclic dispersion in spark ignition engines with high EGR levels. *Int. J. Gen. Syst.* 38, 45-72.

Wang, J. H., Chen, H., Liu, B., Huang, Z. H., (2008). Study of cycle-by-cycle variations of a spark ignition engine fueled with natural gas–hydrogen blends. *Int. J. Hydrogen Energy* 33, 4876–4883.

Wang, J., Huang, Z., Liu, B., Wang, X., (2009). Simulation of combustion in spark-ignition engine fuelled with natural gas-hydrogen blends combined with EGR. *Frontiers Energy Power Eng. China* 3, 204-211.

Yoshioka, Y., Kondo, H., Tabata, Y., Hatakenaka, H., Nakada, K., (2010). Automatic control method of removal of NO by combination of ozone injection and exhaust gas recirculation. *IEEE Trans. Industry Appl.* 46, 1166-1174.

Young, M. B., (1981). Cyclic dispersion in a homogeneous charge spark ignition engine: A literature survey. *SAE paper No. 810020.*

Zervas, E., (2004). Correlations between cycle-to-cycle variations and combustion parameters of a spark ignition engine. *Appl. Therm. Eng.* 24, 2073-2081.

Zheng, J., Wang, J. H., Wang, B., Huang, Z., (2009a). Effect of compression ratio on the performance and combustion of a direct-injection natural gas engine. *Proc. Inst. Mech. Engrs Part D: J. Auto. Eng.* 223, 85-98.

Zheng, J., Huang, Z., Wang, J., Wang, B., Ning, D., Zhang, Y., (2009b). Effect of compression ratio on cycle-by-cycle variation in a natural gas direct-injection engine. *Energy Fuels* 23, 5357-5366.

In: Natural Gas and Petroleum
Editor: Constantin Cranganu
ISBN: 978-1-62808-535-8

Chapter 8

Pore Pressure Prediction Using Velocity-Based Methods

Hamidreza Soleymani[1,*] and Seyedmohsen Seyedali[2]
[1]City University of New York, New York, US
[2]Dana Geophysics Co. Tehran, Iran

Abstract

Over the past two decades, pore pressure prediction represents a key concept in hydrocarbon exploration and drilling. Encountering abnormal pore pressure while drilling is a common issue and has been reported by many drillers worldwide. Pressure kicks, borehole instability problems, stuck pipe, lost circulation, and even blowouts are typical occurring troubles while drilling formations with abnormal pore pressure. Therefore, a relatively reliable quantitative estimation of the pore pressure prior to drilling become imperative for a safe and economic drilling of new wells, especially in formations with potential overpressure risk.

The aim of this chapter is to provide reader with introductory topics in pore pressure prediction. It has been tried to accomplish this goal by avoiding complicated notations and advanced mathematics while, delivering accurate definitions for common petrophysical terms and methods used in pore-pressure prediction literature.

Historically, many geologists tried to classify and explore the geological mechanisms that may involve in producing high pore-pressure (e.g., Fertl, 1976; Fertl and Chapman, 1994). Those efforts revealed that there are number of geologic processes which are commonly culprit in producing overpressure zones. There exist many other complex geologic systems where common proposed origins cannot fully address the true mechanism behind the observed pressure anomalies. However, number of novel models developed by frontier researchers to explain the observed anomalies within those reservoirs (see Deming et al. 2002).

Generally, abnormal pore pressure is a product of geological, physical, chemical and mechanical processes.

---

* Corresponding author: Hamidreza Soleymani. Adress: 2900 Bedford Avenue, Brooklyn, NY 11210. E-mail: hsoleymani@gc.cuny.edu.

Petrophysical and geological studies of various over pressured reservoirs from different part of the world confirmed that usually several pressure producing agents act together in order to generate overpressure within a reservoir.

Currently, geophysical methods especially reflection seismic techniques accompanied by non-seismic methods form a robust tool in pore-pressure estimation. Considering the fact that the fluid pressure has significant effect on compressional wave velocity, analyzing rock velocities became industry standard approach and its reliability confirmed by many geologists and drillers worldwide (Sayers, 2006).

The following chapter contains four sections. Common petrophysical terms involved in pore-pressure studies will be introduced in the first section. Major geological overpressure origins will be discussed in section two. Section three covers the common accepted methods and relationships between different petrophysical parameters. Since rock velocities is the key parameter in many pore pressure estimation methods, understanding the nature of various velocity sources (e.g., velocities from seismic data both pre-stack and post-stack, VSP, logging, laboratory measurements, etc.) can improve our insight and reduce the associated pitfalls in using them. Accordingly, common velocity sources and their applications will be covered as a part of section three. Finally, in the last section of this chapter (section four) a full discussion on using velocity based methods in 3-D pore pressure estimation study provided as a practical example of using velocity based methods in pore pressure prediction studies.

Keywords: Pore pressure-Rock velocity- Seismic velocity- Bowers Method- Sonic log-Geostatistics

INTRODUCTION

One of the key tasks prior to drilling new wells in regions with history of abnormal pore pressure is to evaluate formations with potential overpressure using available data. Drilling in those formations without prior knowledge of pressure distribution could result in losing financial or even human resources. One of the historical disasters happened in Mary Sudik #1 well, drilled by the Indian Territory Illuminating Oil Company (ITIO) in Oklahoma on March 26, 1930. Another recent tragedy happened in 2010 at Deepwater Horizon rig while drilling a well in approximately 1.5 km (5,000 feet) of water in the Gulf of Mexico.

These unpredictable pressure anomalies could be anticipated by study of the pressure attributes using surface (e.g., 2-D and 3D) and down-hole seismic (e.g., VSP) operations, drilling information, and wire logging data (e.g., sonic).

One of the essential studies in order to guaranty the safety and efficiency of the mud weight program is to observe the local fracture pressure gradients and pore-pressure variations in different formations. Also, real time observation of the pore-pressure during drilling phase provides valuable information regarding pressure distribution patterns within or outside the reservoir.

It's been believed by many geologists that undercompaction process is the main origin of the abnormal pore pressure in many part of the world. Accordingly, many empirical relations have been suggested by various researchers to evaluate overpressure due to undercompaction.

Wire log data considered as one of the most reliable and valuable sources of data which contains useful information about formation pore-pressure.

In one of the classical papers published by Fertl (1976) he discussed the importance of the wire logs and the way they should be used for pore-pressure prediction assessment.

Other valuable sources of information which are usually acquired during the exploration and evaluation phase or while production of the oil and gas reservoirs are reflection seismic data. Without doubt the most important contribution of the seismic data are inferring basin stratigraphy and major formation horizons which could be derived directly from the post stack seismic data. This information is the key input in building geologic model of the reservoir.

Gregory (1977) showed that different pore fluids with various pressures have a significant effect on rock velocities. He established these results base on the laboratory measurements of different elastic waves (shear and compressional waves) travel time in various rocks, fluids and pore-pressures. Base on similar petrophysical researches many scientists tried to propose a reliable relationship between the pore pressure and rock velocity.

Obtaining relatively accurate velocity data is still subject of major debate. Having reliable relation between rock velocity and pore pressure without consistent rock velocities will result in imprecise pressure estimations.

Typically, geophysical pore pressure prediction methods are based on the effects of pressure on seismic wave's velocity (especially P-wave velocity). It is supported with many written documents that the geophysical methods are capable of reliable estimations of the pore pressure (see e.g., Cranganu, 2007; Bowers, 2002; Sayers, 2006; Soleymani and Riahi, 2012). Following the pioneer work done by the Pennebaker (1968), Dutta (1987) published a comprehensive paper wherein, he introduced and discussed various geophysical methods of pore pressure prediction study. Today, high-resolution post and pre stack 3-D seismic data improve the results of the predictions tremendously.

One of the valuable information that could be derived directly during processing of the 2-D and 3-D reflection seismic data are stacking velocities. Stacking velocities defined as the distance-time relationship determined from analysis of normal moveout (NMO) measurements from common depth point gathers. Online oil field glossary of Schlumberger defined stacking velocity as following: "The stacking velocity is used to correct the arrival times of events in the traces for their varying offsets prior to summing, or stacking, the traces to improve the signal-to-noise ratio of the data". Stacking velocities could be converted to interval velocities through applying Dix conversion (Dix, 1955).

On top of all benefits of seismic data in evaluation and exploration of the hydrocarbon reservoirs, offshore seismic enables drillers to examine seabed and its properties prior to drilling. This will result in substantial improvement in offshore drilling efficiency.

Seismic data enable us to do the following important assessments:

1. Abnormal pore pressure evaluation using velocity data prior to drilling new wells.
2. Study of the fluid mobility and dynamic pressure changes using 4-D seismic.

Dutta (2002) stated that almost all quantitative methods of overpressure prediction take advantage of the following properties:

- Relatively low bulk density
- Relatively low seismic velocity

- Relatively high porosity
- Relatively decrease in effective stress
- Relatively high formation temperature
- Relatively high poison ratio
- Relatively low interval velocity

As a general rule, rock velocities obtained from sonic log and seismic data or from any other sources of the velocity, tend to increase under normal compaction regime. Each major deviation from this trend could be considered as abnormal pore pressure or existence of other anomalies like gas saturation.

In many sedimentary basins the measured pore-pressure at a certain depth could be either higher (overpressure) or lower (underpressure) than the normal (hydrostatic) fluid pressure at that depth. Sedimentary basins experiencing one or both of these situations is said to have abnormal pressure regimes. Overpressured zones are far more common than underpressured zones (Cranganu, 2007). Abnormal pressure regimes are sources of the considerable variations of velocity with increasing depth.

These pressure variations can also have different origins (e.g., petrology, hydrocarbon saturation, formation temperature). One of the main purposes of pore pressure study is to observe and clarify those velocity anomalies associated with different sources.

1. Pore Pressure Basics

There are several common geomechanical terms and definitions involved in pore pressure study. Almost every published paper on this subject begins with defining set of common terms.

Accordingly, in this chapter we tried to provide the simple and accurate definitions for the most important geomechanical terms that appear almost in every pressure related paper. Here we tried to do this by avoiding advanced mathematics and heavy notations without trading simplicity over accuracy.

The main goal of the pore pressure estimation is to obtain safe and economic mud weight program. Considering the importance of mud weight program and its link to pore pressure prediction, some basic terminology included in this chapter to familiarize reader with the essential notions concerning drilling and mud weight planning.

1.1. Hydrostatic Pressure

Hydrostatic pressure or normal pressure is referred to stress exerted by the weight of the static fluid column. Hydrostatic pressure is a function of the height and density of the pore fluid and it is independent of the geometry of the fluid column. Density of the fluid is function of the fluid type, aggregation of the unsolved solids (i.e. salt and other minerals), existence of the gases, temperature, and pressure.

Similar to rock matrix, density of the fluids tends to increases with depth. Here depth refers as the vertical distance between the measured point and the reference datum.

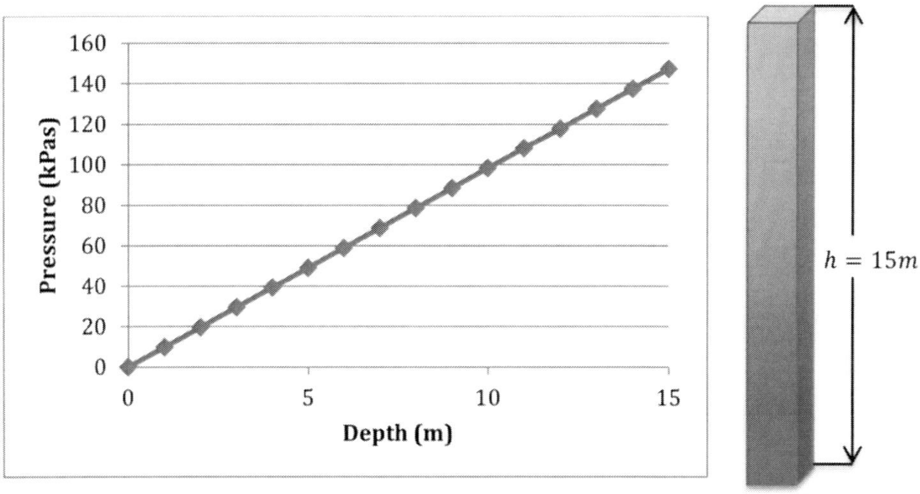

Modified after Brown, 1967.

Figure 1. Schematic plot of the hydrostatic pressure for 15m height pure water. The slop of the line is (pressure gradient) is 9.82 kPa/m.

From the plot of the pressure versus depth for pure water it can be inferred that the slop of the gradient line would be 9.82 kPa/m (Figure 1). If we draw the same line for the formation water, usually the slop of the line will vary between 10.18 -10.40 kPa/m.

Clearly, the observed difference is due to higher density of the formation water in compare with pure water. In porous rocks, where pore spaces are connected to each other all the way up to the surface (open system), weight of the above fluids creates the observed pore fluid pressure in the lower parts. This pressure is referred as the hydrostatic pressure or normal pressure. Hydrostatic pressure is calculated using following equation:

$$P_{normal} = g\int_{0}^{h} \rho_{fluid}(z)dz \tag{1}$$

where P_{normal} is hydrostatic pressure in Pascal, ρ_{fluid} is the density of the fluid at depth h in kg/m^3, g is the gravitational acceleration in m/s^2 and h is the height of the fluid column in m.

1.2. Overburden Pressure

Overburden pressure is defined as the pressure exerted by the weight of the sediments and their containing pore fluids. This could be calculated using following equation:

$$S = g\int_0^h \rho_{bulk}(z)dz \tag{2}$$

again, g is gravitational acceleration in m/s^2 and h is the vertical height of the sediment column in m, ρ_{bulk} is bulk density and it's calculated using following equation:

$$\rho_{bulk} = \varphi\rho_f + (1-\varphi)\rho_m \tag{3}$$

where φ is porosity of the sediments and its value varies between 0 and 1, ρ_f is formation fluid density in kg/m^3, ρ_m is rock matrix density in kg/m^3.

Generally, porosity of the sediments decreases with increasing depth (compaction process). Decreasing porosity will result in increasing density. As a result of the compaction process, overburden pressure increases with depth.

1.3. Pore Pressure

Simply, pore pressure is the pressure of the fluids inside the rock pores which presented by the pressure units. It can be normal, overpressure or underpressure.

Normal (hydrostatic) pore pressure is referred to the pressure of the pore fluids with expected hydrostatic pressure at particular depth. In contrast, Underpressure is referred to the state where pore fluid pressure is less than the expected hydrostatic pressure at certain depth. Finally, overpressure is referred to the situation where pore pressure exceeds hydrostatic pressure. Theoretically, overpressure is confined by overburden pressure.

1.4. Effective Pressure

Effective or differential pressure refers to the pressure exerted on the solid frame of the rock. Terzaghi and Peck (1948) proposed that the elastic wave velocity is the function of pore pressure and the total stress tensor S_{ij}. They introduced the following relation:

$$\sigma_{ij} = S_{ij} - \alpha p \delta_{ij} \tag{4}$$

where $\delta_{ij} = 1$ if i=j and it is 0 otherwise. The parameter σ_{ij} is called effective stress tensor. The parameter α is constant and according to Eaton (1972) and Bowers's (1995) assumption is equal to 1. With this assumption, we can rewrite the Terzaghi's effective stress as following (Equation 4):

$$\sigma_{ij} = S_{ij} - p\delta_{ij} \tag{5}$$

σ_{ij} is referred as the differential stress tensor or effective stress. The effective pressure is the pressure which is acting on the solid frame of rock. Normally, in sedimentary basins, elastic wave velocity rises with increase in depth.

In uniaxial stress case, it is assumed that velocity and porosity of sediments depends only on vertical component of stress. By applying this assumption we can rewrite equation 5 as following:

$$\sigma = S - P \tag{6}$$

where σ and S are vertical components of S_{ij}, σ_{ij}. Figure 2 is the simple schematic representation of equation 6.

Tosaya (1982) showed that the effective pressure is the key controlling factor in compaction process. This statement is even more pronounced in sedimentary rocks. Also, he showed that the decrease in the effective pressure will lead to decrease in compaction rate. Consequently, abnormal pore pressure will produced within the formation. In overpressure formations, part of the weight of the overburden rocks and sediments carried by the pore fluids. As a general rule existence of the overpressure and/or high porosity will reduce the acoustic wave velocity. Figure 3 is an integrated schematic sketch of the hydrostatic pressure, overburden pressure, normal, and overpressure as a function of the burial depth.

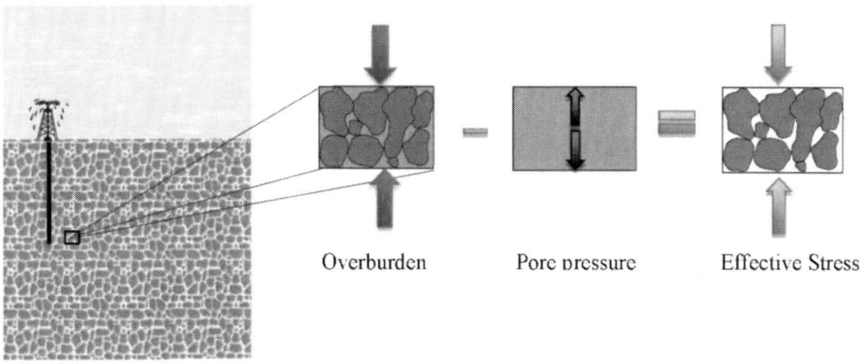

Modified after Tosaya, 1982.

Figure 2. Schematic illustration of equation 6. Effective stress is the pressure exerted by the weight of the above sediments on a solid rock frame. This stress can be calculated using Terzaghi's effective stress equation presented by equation 6.

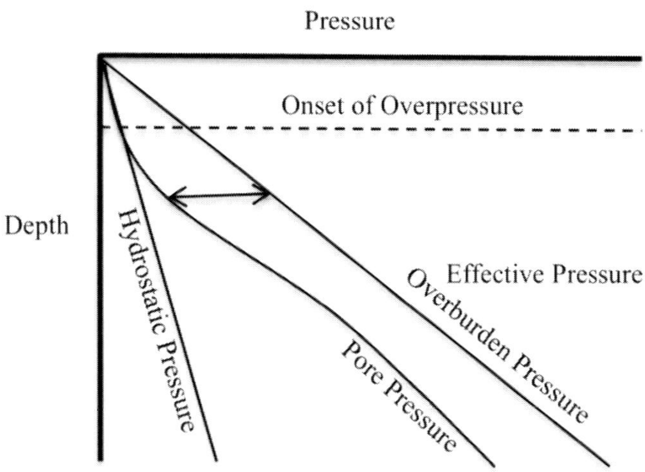

Figure 3. Integrated schematic sketch of the hydrostatic pressure, overburden pressure, normal and overpressure. According to Terzaghi and Peck (1948) effective pressure produced by differentiating overburden pressure from pore pressure.

1.5. Pressure Gradients and Mud Weight Program

Pressure gradient is a common used term in drilling industry. For the sake of simplicity, drillers prefer using gradients in their calculations instead of correct pressure units. Pressure gradient is defined by increase in pressure with increase of 10 m in depth. Respectively, overburden and fracture gradients defined as their raise by increasing 10 m in depth.

One of the most important gradients is the Equivalent Mud Weight or simply EMW. EMW is expressed using following equation:

$$EMW = PRESSURE \times \left(\frac{1}{0.052}\right) \times \left(\frac{1}{TVD}\right) \tag{7}$$

Typically EMW expressed in pound per US gallon (PPG); pressure in PSI and the total vertical depth (TVD) is expressed in feet. The constant value of 0.052 is a conversion constant that can be derived by dimensional analysis. Using metric units, pressure expressed in Pascal and TVD in meter will lead to conversion constant equal to1175.7 (unit less).

However, it should be noted that using density unit to describe pressure is absolutely wrong but drillers prefer using this unit since it gives them more sense about drilling difficulty. For instance knowing the formation pore pressure is 4500 psi (31026.4 kPa) do not implies that whether this formation is easy to drill or not. In 10000 ft (304.8 m) this pressure is equal to hydrostatic pressure, but in 500 ft (152 m) this pressure considered as a substantial pore pressure. Let's clarify this problem by converting and comparing pressures to EMW. At 5000 ft (152 m), EMW= 17.5 lb/gal (260.014 N/m^3) and at 10000ft (304.8 m), EMW=8.65 lb/gal (177.95 N/m^3) (Bruce and Bowers, 2002). As mentioned previously, pore-pressure confined by the fracture pressure. If drilling fluid pressure exceeds fracture pressure, formation rocks will eventually begin to fracture. Notice that the lower and upper bound for drilling fluid pressure is pore pressure and fracture pressure respectively (Figure 4). If drilling fluid pressure exceed the fracture pressure then, it's possible that the mud pressure, fracture the formation. Consequently, mud circulation could be lost and the mud flow through the formation cracks.

Lost circulation considered as one of the most serious problems occurred during drilling new wells. By definition lost circulation referred to the situation that drilling mud flow through the formation instead of circulating back to surface through annulus. The least possible issue associate with lost circulation is to lose money by losing drilling mud. Also, enormous loss of drilling mud could result in dropping hydrostatic pressure exerted by the mud weight. If drilling fluid pressure was less than pore pressure then pore fluids would influx into the wellbore and cause kicks, stuck pipe or even blowout.

Figure 4 is a redraw of the Figure 3 with some extra features. In this Figure fracture pressure is shown by dash line. Theoretically fracture pressure is less than overburden pressure. As mentioned previously, for a safe and optimized drilling, weight of the drilling fluid should be narrowed between fracture pressure and pore pressure. In Figure 4 mud weight shown by solid line.

Generally, drilling begins with low mud weight. As the pore pressure increases and reached almost to mud weight then, drillers increase the mud density to compensate the increase of the pore-pressure in deeper depths.

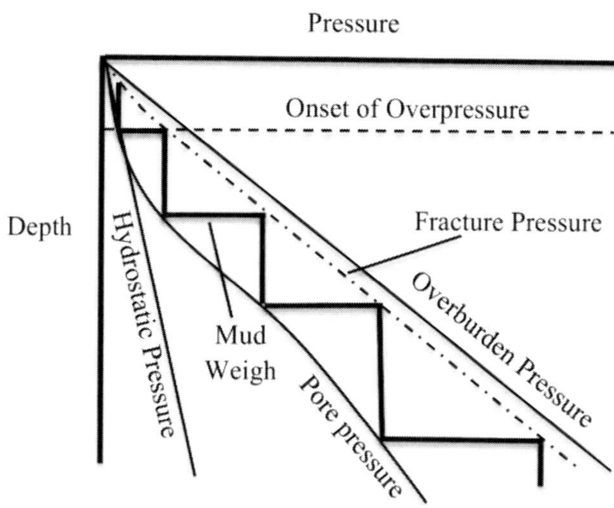

Figure 4. Schematic representation of overburden, effective, fracture, mud weight and hydrostatic pressure. As the Figure implies pore pressure cannot exceed hydrostatic pressure. At top of the overpressure zone pore pressure exceeds hydrostatic pressure.

Obviously, if drillers start with high mud weight the well would fracture immediately. Thus, pervious knowledge or estimation of the pore pressure can play crucial role in efficient, safe and economic mud weight planning.

2. Major Geological Origins of Abnormal Pore Pressure

Many geological processes are associated with overpressure. In fact, geologists believe that; in many cases several factors act concurrently in order to form an overpressure system. Generally speaking, geological, physical, chemical and mechanical processes are the main factors involving in producing overpressure. Pore pressure is a hydrodynamic notion where time plays important role both in its creation and elimination. This means that every overpressure zone has a limited lifespan that bound by existence of its main origin and cap rock resistance toward increasing pressure. One of the necessary elements in presence of overpressure is an isolate or semi-isolate environment. Clay deposits have relatively low permeability. These deposits do not allow fluids to flow out easily through certain time span (geological time scale). Thickness and degree of capillary of Clay deposits are two important factors governing the integrity of the clay formation as a seal. As mentioned before, many sources can associate with abnormal pore pressure, in this section most common geologic origins of overpressure will be discussed briefly.

2.1. Overburden Effect

The principal factor governing overburden effect is the ability of the formation in retarding pore fluids to discharge from the rock pores.

During sediment compaction, the porosity of the sediments decreases as the fluids leaving the pore spaces. Seabed shales can have up to 80% porosity although; this relatively high porosity could be reduced up to 20% at depth 1000 m. Notice that the porosity gradient will decrease in deeper depths. During sedimentation phase, the most important factor controlling pore fluid discharge is increase in overburden weight. The sediments are called normally compacted if the sedimentation rate was slow enough; allowing pore fluids leave the pore spaces. There exists a fine balance between sedimentation rate and pore fluid discharge. Naturally, shales have relatively low permeability. There is a minimum permeability for a particular compaction rate which allows formation fluids to flow easily during sedimentation. This balance of permeability and sedimentation rate will result in normal compaction of the sediments. Either decrease in permeability or increases in sedimentation rate will retard the outflow of the pore fluids which consequently, increase the pore pressure. Rising pore pressure will continue until the pore fluids find a way-out form the trap. Usually, permeability considered as a relatively constant parameter. Thus, increase in pore fluid pressure controlled mainly by sedimentation rate. In general, abnormally pressures zones are most common occurring in active continental margins and young subduction zones. It's important to know, that undercompaction cannot reduce the effective stress.

"Under non-decreasing effective stress condition, plot of the effective pressure versus velocity called virgin curve" (Bowers, 1995). Schematic drawing of the virgin cure is shown in Figure 5. Normally, undercompaction itself, cannot reduce effective pressure thus, virgin curve still governs the relation between velocity and effective pressure. In extreme cases, under compaction freeze the effective stress, which results in holding velocity around some particular point. In Figure 5 velocities drawn versus depth for the normal velocity trend and expected trend in undercompaction zone.

2.2. Fluid Expansion Mechanism

Fluid expansion mechanisms are referred to mechanisms that expand the pore fluid volume.

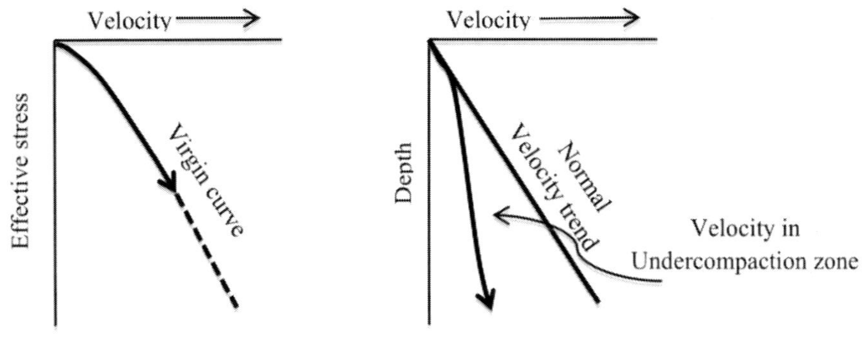

Modified from Bowers, 1995.

Figure 5. Schematic drawing of the virgin cure. Plot of the effective pressure versus velocity in case of no decrease in effective stress, the resulting curve is called virgin curve (left). Also, in this Figure velocity drawn versus depth for normal velocity trend and expected trend in undercompaction zone (right).

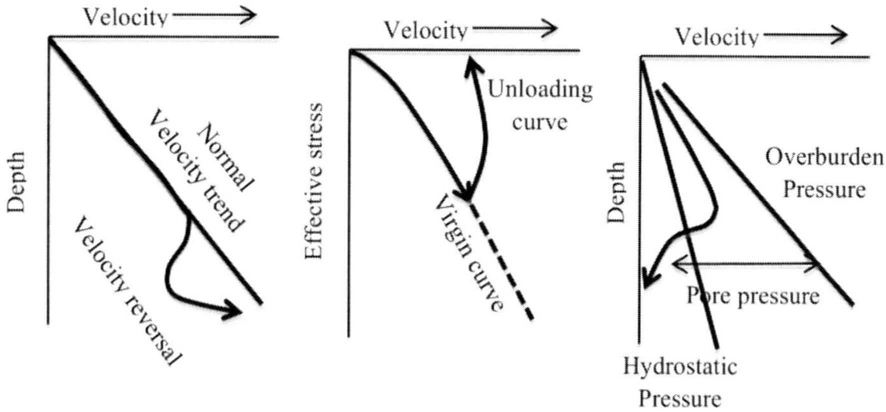

Figure 6. This Figure shows the effect of fluid expansion on velocity and pore pressure trend. Figure in the left shows velocity for normal and overpressure state due to fluid expansion mechanism. Plot in the middle shows unloading curve introduced as the result of fluid expansion mechanism. Finally in the right Figure, velocity plotted versus depth for the normal compaction, overburden pressure, undercompaction and finally overpressure due to fluid expansion mechanism.

Comparing pore-pressure creation ability associated with undercompaction and fluid expansion, the later can increase pore pressure with much higher rate. Fluid expansion mechanisms are capable of decreasing the effective pressure (unloading effect), which will cause velocity reversal. Most people associate unloading with loss of overburden by erosion, which is not correct. Comparing unloading with hydraulic jack; "pumped oil in jack reduces the load that the car's springs carry" said Bowers (2002). In a similar way, expanded fluids in rock pores will reduce the load exerted on rock frame. Figure 6 shows the effect of fluid expansion on velocity and pore pressure trend. Figure in the left shows velocity for normal and overpressure state due to fluid expansion mechanism. Figure in the middle, shows the unloading curve as produced by fluid expansion mechanism. Finally in the right Figure, velocity plotted versus depth for the normal compaction, overburden pressure, undercompaction and finally overpressure due to fluid expansion mechanism.

Generally three mechanisms are considered as the common main origins of the pore fluid expansion:

- Expansion as result of heat
- Diagenesis
- Hydrocarbon creation and oil-gas conversion

Heat can impose a relatively great deal of pressure to formation fluids. Generally, fluids tend to expand more than solid rocks frame. Also, heat can reduce the viscosity of the fluids. Consequently, low viscose fluids can move much easier. According to Magara (1975) fluid expansion is only about 0.05 percent in burial depth of 1km and temperature gradient of 25 °C /km. Considering this tiny expansion and viscosity reduction, integrity of the cap rock become an important factor in producing overpressure. Therefore, Small fractures in cap rock can discharge the pressure produced by heat. Diagenesis is referred to any physical, chemical changes undergo by sediment after lithification.

Three major stages included in diagenesis; producing newborn minerals, rearrangement or recrystallizing of minerals in sediments and lithification.

Clay diagenesis is believed by many scientists that have an important role in producing overpressure. Various studies on clay minerals showed that by increasing depth smectite layers transform to illite. Smectite illite transformation controlled by seabed temperature, geothermal gradient, and sedimentation rate (Figure 7).

Intergranular water releases from smectite during Smectite illite transformation. This additional water increase pore fluid volume which in return can increase pore pressure. Diagenesis considered as a major cause of abnormal pore pressure in Gulf of Mexico, West Africa and south East Asia.

Evaporite deposits can also have two distinctive roles in producing overpressure zones. Since they are impermeable, they can form a good cap-rock. Also, these deposits can fill their small fractures and form resistant cap-rock which has the ability of preserving high pore fluid pressure. This ability is more distinct in salts.

Evaporite deposits can also, increase pore fluids through diagenesis process. The process of transforming gypsum to anhydrite produces water as high volume as 35 percent of the initial gypsum. This excess in pore fluids can produce extra pressure in formations with good and impermeable cap-rock. Anhydride, hydrates through similar process which will lead in increase of its initial volume. These processes cannot act as primary causes of abnormal pore pressure but they can only intensify fluid pressure within the formation.

Bacterial activity in shallow depths can produce biologic methane from organic matter breakdown. Since there is low chance of existence of a good cap-rock in shallow depths, these gas pockets usually find their way up to surface. High resolution seismic data could be very useful in detecting these gas pockets since they are situated in shallow depths and seismic data has much finer resolution in near surface travel times.

Bacterial activities are low in relatively deep depths. Alternatively, thermal breakdown initiated at those depths. During thermal breakdown process, heavy oil molecules break down and produce much lighter products.

This process will result in increase in number of molecules which in return expand the fluid volume. Thermal breakdown considered as one of the secondary causes of overpressure in close or semi-close environments.

2.3. Tectonic Process

Tectonic process can either produce pressure anomalies or return the system to its normal state. Similarly faults can form a seal or a drainage system. Normal faults are formed by movement of the fault blocks away from each other (tensile mechanism).

Normal faults are usually forming a drainage system where they connect different parts of reservoir. In certain situations and in presence of the saturated fluids, pressure decrease around the fault plain can form a proper place for crystallization of calcite, quartz and dolomite. This process could seal the reservoir.

Unlike the normal faults, Trust and reverse faults are tending to close since they produced as a result of compressional stress. Strike slip faults are also acting similar to normal faults and they can either form drainage or seal system. Normal faults and there role in producing abnormal pressure showed in Figure 8.

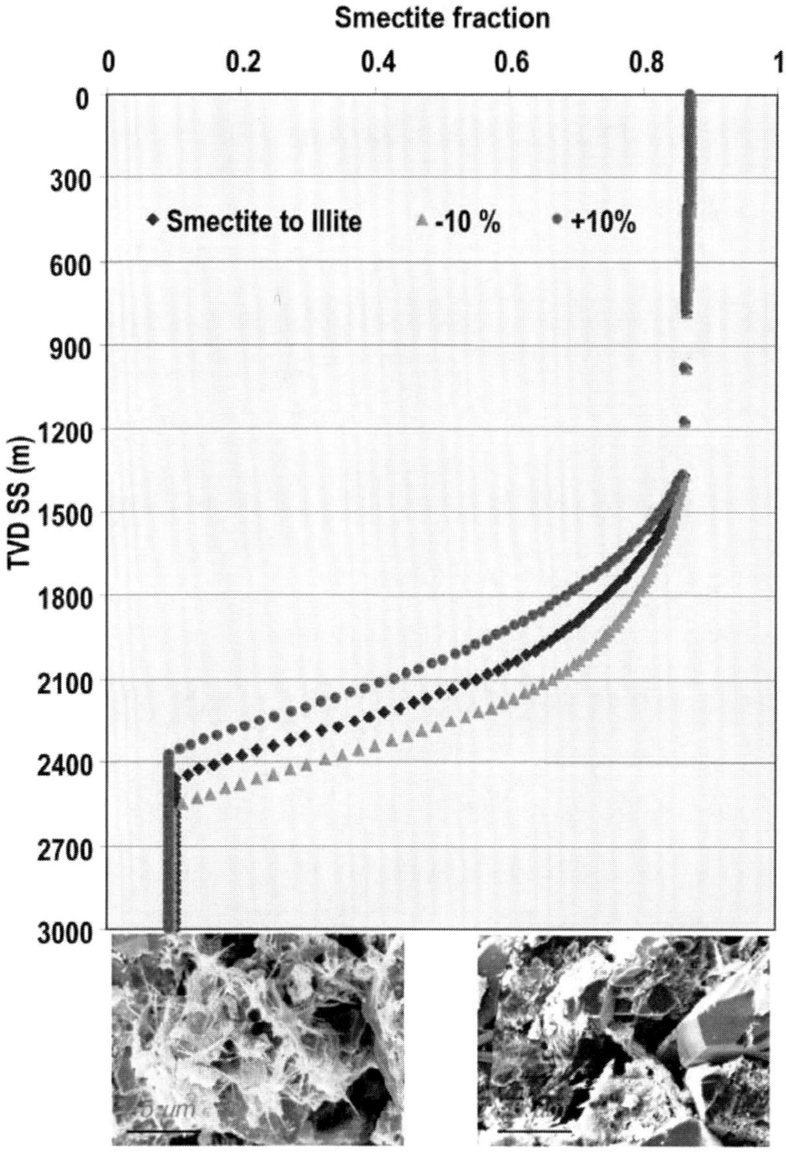

Figure 7. "This Figure shows a digenetic profile of smectite rich rock transforms into illite as a function of the depth. Geothermal gradient is very important in diagenesis changes. Here changes in smectite to illite transform shown with ±10 percent change in geothermal gradient kinetic parameters for the transition are obtained by data fit to X-ray diffraction data on mixed-layer illite/smectite clays from fine-grained (less than 5 micron) sediments from many wells. The pictures at the bottom are the electron microscope images of a smectite-rich (right) and illite-rich (left) sandstone" (Dutta and Khazanehdari, 2006).

Fractures are cracks in rocks with no or little displacement. They play important role in integrity of the cap-rock. Plastic clays, anhydrite and salt depositions are capable of regaining their integrity. This behavior can be seen frequently in Iran salt deposits. Figure 9 shows a typical salty formations of Iran's south west on seismic data.

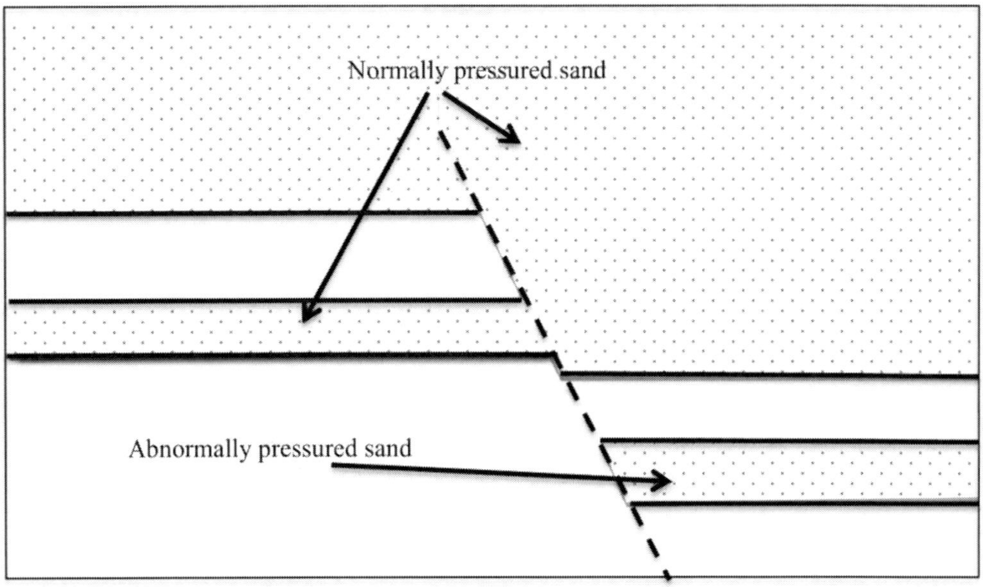

Figure 8. Normal faults and there role in producing overpressure zone.

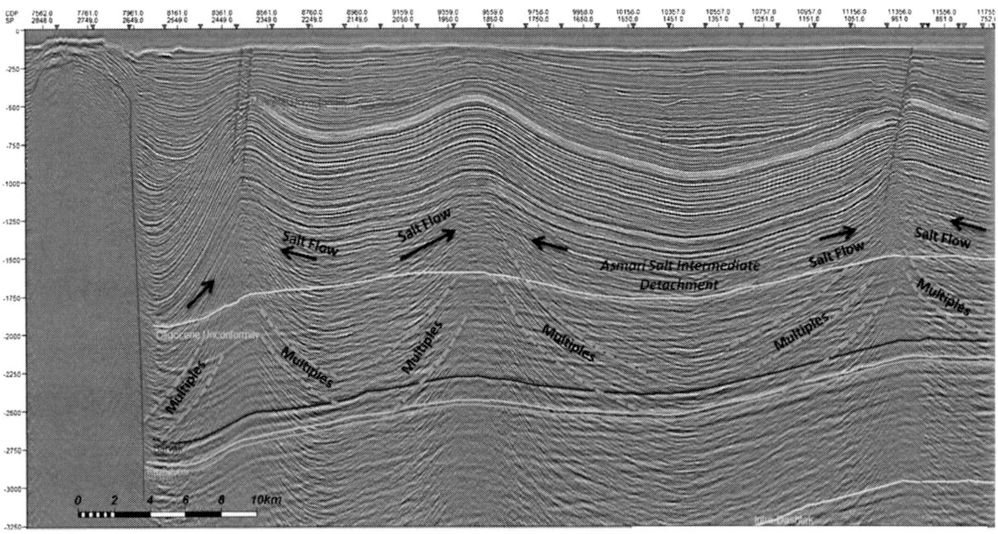

Figure 9. Typical salty formations of Iran's south west on seismic data. Plastic clays, anhydrite and salt depositions are capable of regaining their integrity. This behavior can be seen frequently in Iran's salt deposits.

3. Petrophysical Relations, Velocity Sources and Pore Pressure Prediction Methods

It's been cited by many authors that geophysical methods in conjugate with other techniques can form a reliable tool for pore pressure prediction purposes (Bowers, 2002; Sayers, 2006). Overpressure shales considered as a good cap-rock.

On the other hand, these formations could be troublesome for drillers. Drillers in those formations should always consider proper safety margin in their mud weight program. Penebaker is one of the pioneers in quantitative estimation of pore-pressure. He tried to estimate pore pressure using sacking velocities (Pennebaker, 1968).

Another classic but noteworthy publication wrote by Dutta (1987) wherein; he discusses methods of geopressure prediction. Today with advent of high resolution 3-D and three component seismic data the precision of the estimations improved tremendously.

Generally it's possible to categories measurements and observations in different development stages of the new wells as following:

Prior to drilling, it's possible to use surface geophysical methods (e.g., gravity data, two and three-dimensional seismic, three component seismic data, velocity of compressional and shear waves, density and porosity data) to evaluate pore pressure.

It's feasible to evaluate pressure regimes in new wells by using acquired information (e.g., drilling parameters, shale cuttings and drilling mud weight) during drilling stage of offset wells.

Other information namely, drilling penetration rate, logging while drilling, seismic while drilling could be valuable extra information but they are not always available. Also parameters like pressure kicks, mudflow rate, formation factor, density, volume of the shale could be used to assess pressure regime during the drilling stage.

In the testing and competition stage of the new well, it's possible to study pressure in small zones using tests e.g., repeat formation test, pressure bombs.

In normally compacted settings (in absence of hydrocarbon saturation) sonic velocity tends to rise with increasing depth. Each unexpected deviation from this trend considered as abnormal pore pressure.

This section deals with the basic relations between different petrophysical parameters. Since study of the rock velocity is the base of many pore pressure methods, a separate discussion included in this section to address different velocity sources. Finally industry accepted methods of pore pressure prediction will be discussed with required detail.

3.1. Relationship between Porosity, Velocity and Effective Pressure

Obviously, compaction process will result in reducing porosity. Therefore, porosity could be considerate as one of the main signatures of the compaction. Athy (1930) relates porosity to depth by the following relationship:

$$\varphi(z) = \varphi_0 e^{-cz} \tag{8}$$

In this equation, $\varphi(z)$ is the porosity at depth z, φ_0 porosity at z=0 and c is the environmental constant. From 1930 up to now, many authors suggest different equations. One of these relations proposed by Stump et al. (1998):

$$\varphi(z) = \varphi_0 e^{-k\sigma} \tag{9}$$

In this equation, k is the density of the sediments and their pore fluids and σ is the effective pressure. Equation (9) could be rewritten by taking logarithm form both sides. The logarithm format will ease the calibration and consequently, using the equation.

$$\ln\varphi = \ln\varphi_0 - k\sigma \tag{10}$$

Other scientists suggest different relations. For instance Terzaghi and Peck (1948) suggested following relation between porosity and effective pressure:

$$\varphi = 1 - \varphi_0\sigma^{C/4.606} \tag{11}$$

Also, Dutta (1987) suggest an empirical relationship for calculating effective pressure as a function of temperature:

$$\sigma = \sigma_0 e^{\varepsilon\beta(T)} \tag{12}$$

where $\varepsilon=\varphi/(1-\varphi)$ and $\beta(T)$ is the function of the temperature. σ_0 is a constant value and depends on dominant lithology. The calculated effective pressure could be converted to pore-pressure using simplified Terzaghi equation 6. Many geoscientists tried to provide a comprehensive equation governing relation between velocity and porosity. Although those efforts result in many reliable imperial relations, studies still continue to find dependable equations especially for shales.

One of the early relationships between porosity and transient time is given by Pickett, (1963):

$$\Delta t = A_0 + B_0\varphi \tag{13}$$

In which Δt is transient time in ms, A_0 and B_0 are lithology dependent constants. Other widely acceptable equation between density and velocity proposed by Gardner and Gregory (1974). They relate velocity to density using following equation:

$$\rho = \alpha V_P^{\beta} \tag{14}$$

where ρ is density and V_p is P wave velocity.

Equation (14) could be calibrated in order to find α and β with known velocity and density values at well locations.

In 1995, Bowers suggested an outstanding empirical equation, using data from Gulf of Mexico, which governs the compaction and fluid expansion mechanisms. Since these two mechanisms are the most frequent causes of overpressure, the Bowers's equation 15 is an appropriate approach and it is applicable in most cases. In order to obtain desired accuracy Bowers's equation (Equation 15) should be calibrated by calculating constants A and B (i.e. the virgin curve parameters) using available data from offset wells in particular field.

$$\sigma = (\frac{V - 5000}{A})^{\frac{1}{B}} \tag{15}$$

here V is velocity in (ft/s), σ is effective stress in (psi). At well locations with the known pressure, velocity and density values one can easily calculate the A and B (constants of Bowers's equation 15). Note that this equation should be used for calculating effective pressure with known velocity within the virgin zone (in the absence of unloading mechanism).

In existence of unloading general Bowers's equation (Equation 16) should be used.

$$\sigma = \sigma_{max}^{(1-U)}(\frac{V - 5000}{A})^{\frac{1}{B}} \tag{16}$$

where σ_{max} is defined by:

$$\sigma_{max} = (\frac{V_{max} - 5000}{A})^{\frac{1}{B}} \tag{17}$$

According to Bowers (1995) V_{max} and σ_{max} are estimates of the effective stress and velocity at the onset of unloading.

3.2. Common Velocity Sources for Pressure Prediction Study

As previously mentioned, rock velocity data could be converted to effective pressure using proper equation. Effective pressure is key data in calculating pore pressure. In recent years, numerous techniques developed to extract acoustic wave velocities from raw seismic data with required resolution and accuracy for pore-pressure estimation study (see Chopra and Huffman, 2006). Aside from derived velocities from seismic data, other sources of velocity e.g., velocities obtained from lab study, sonic logs and vertical seismic profile (VSP) data, are also considered as reliable sources of velocity information. Basic understanding of the different velocity sources will prevent us from common pitfalls associated in using different velocity data. Sonic logging device measures rock velocity by recording transient time between sent sonic pulses in one depth and received pulse at other depth inside the well. By having transient time and space length between receiver and transmitter it is easy to calculate interval rock velocity. Since calculations and measurements are along the space between transmitter and receiver, acquired velocity data form sonic log will be interval velocity. These velocities will be different from velocities data obtained from core analysis in lab. Usually in Sonic logging device space between sonic source and receiver is about 0.60m. Thus, this small space promises a good match between sonic velocities from logging and lab observations. VSP (vertical seismic profiling) data is another reliable source of velocity information. As its name would suggest, in this method seismic source located near the well head and receivers set down the well.

Classical VSP data acquired by a repetitive seismic source like Vibroseis and a moving receiver inside the well. By each pulse of the seismic source the receiver moves commonly 15m (the interval varies upon the required resolution).

Here the transient time between two shots divided by the distance between source and receiver for two adjacent records, the result will be the interval velocity between two receiver locations. The acquired velocity data from VSP is much or less the same in compare with velocities obtained from sonic log. The nature of disparity between VSP interval velocity and sonic log comes from the frequency content differences. Sonic log frequency content is 10-20kHz while VSP is 10-125Hz.

Another source of disparity in interval velocity between log and VSP data is fact that the lithology changes between two adjacent receiver positions in compare with the common 0.60m space between receiver and transmitter in sonic log tool. Since VSP data uses down-hole seismic tool and records one way transient time, the velocity data from VSP is more accurate than the velocity calculated using 2-D or 3-D seismic. Due to finer resolution and smaller receiver transmitter space in the sonic log, this device is more exposed to environmental factors in compare with VSP data. Resulting velocities from VSP data could be used to correct seismic derived velocities.

In 1968, Pennebaker published a paper which considered among the first publications in pore pressure study. In that paper, he used stacking velocities as a main source of velocity data. Stacking velocities is the mid-processing product of reflection seismic data processing. Stacking velocities calculated from analysis of the travel time equations for a horizontal layered earth model. Figure 10 shows a horizontally n layered earth model and reflected ray path from n^{th} layer:

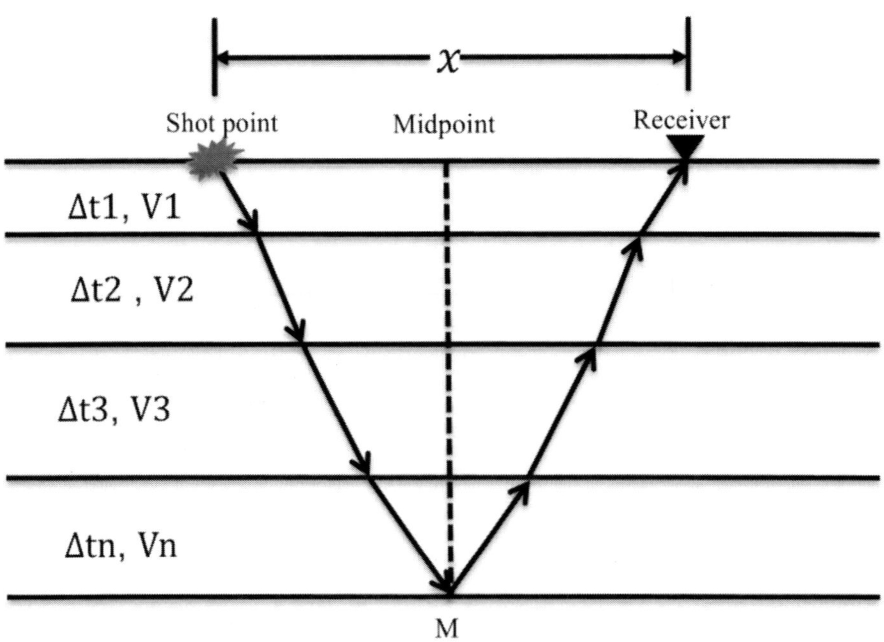

Figure 10. Schematic view of the horizontally n layered earth model and reflected ray path from layer n. In this Figure x represents a distance between shot point and receiver.

The travel time for a single ray-path calculated using Tanner's equation (Tanner and Koehler, 1969):

$$t^2(x) = C_0 + C_1x^2 + C_2x^2 + \ldots \tag{18}$$

where t(x) and t(0) is travel time in offset x and x=0 respectively. C_0 and C_1 are:

$$C_0 = t^2(0) \quad , \quad C_1 = \frac{1}{V_{rms}^2} \tag{19}$$

here C_2, C_3 and … are terms that are function of layer thickness and interval velocity.

RMS velocity could be calculated using following equation:

$$V_{rms}^2 = t(0)^{-1}\sum_{i=1}^{N} V_i^2 \Delta t_i(0) \tag{20}$$

It's possible to eliminate the terms that came after C_1x^2 in equation 18. This is the valid simplification by comparing the distance x to depth. Equation 18 could be rewritten as following:

$$t^2(x) = t^2(0) + \frac{x^2}{V_{rms}^2} \tag{21}$$

Comparing equation 21 with well-known equation for normal move out correction (NMO) (Equation 22) for the horizontal layer earth model; it can be easily concluded that V_{nmo} is equal V_{rms} by some approximations. It is important to know that the normal move out velocity concept is different from stacking velocity, since stacking velocity is the velocity which allows the best stacking of common midpoint gathers (CMP).

$$t^2(x) = t^2(0) + \frac{x^2}{V_{nmo}^2} \tag{22}$$

Chopra and Huffman (2006) published a paper in which he tried to introduce different velocity sources suitable for pore pressure prediction study. In his paper, he tried to address the ambiguity in using different velocity sources.

4. Practical Example of Velocity Based Pore Pressure Prediction

In this section we tried to show the practical application of what was said in past sections by reviewing a real world pore-pressure project.

In this example we tried to generate pore-pressure using different available velocity sources (stacking, VSP, and sonic log velocities). This section is an extended review of the pore pressure prediction project done by Soleymani and Riahi. (2012). Consequently, the author reused the content of the aforementioned paper extensively.

4.1. Pore Pressure Prediction at One of the Iranian Oil Fields

The studied oil field is located in the southwest of Iran. This field does not have any outcrop and explored during a seismic exploration in 1962. The first exploration well which was drilled in 1963 confirmed the existence of hydrocarbon in the region. In this field the length of oil water contact is approximately 39 km, and its width is about 3.5 km. It is worth mentioning that this structure has similar bearing as the Zagros mountain (NW-SE). Totally 60 wells are drilled in this field so far and the data from 35 wells are used in the presented study. The initial aim of this study within the framework of the development project is to evaluate the pore pressure and detect possible pressure anomalies within the productive reservoir.

One of the important steps in pore pressure estimation based on velocity data is to generate an accurate velocity model with suitable resolution for well planning purposes. In this study, acoustic velocities from well logs and seismic interval velocities obtained from stacking velocities are employed along with acoustic impedance inversion of post-stack seismic data to build a valid velocity model. Figure 11 shows the whole workflow of this study. Since different parts of the model have different available data, the velocity model split vertically into two segments.

The first segment, ranges from datum to upper edge of the horizon "A" which is a sharp reflector observed approximately in 1350ms within 3-D seismic cube. The only available data for the first segment is interval velocities obtained from staking of 3D seismic data and zero offset VSP at 2 well locations.

The second segment continues from horizon "A" to horizon "B". Figure 12 shows the horizon A and B on the 3-D seismic data (above Figure) while these two horizons shown in 2-D section in bottom Figure. The interval between horizon A and B constitutes our target reservoir. It should be noted that all available logging data are acquired within the reservoir zone (second segment). The final aim of this study is to estimate reservoir's 3-D fluid pressure.

Considering the numerous wells within the reservoir which contains sonic and density logs, it is decided to horizontally interpolate the derived velocities form sonic logs, by applying geostatistical methods such as co-kriging technique. The secondary information used in this procedure is acoustic impedances from post stack-seismic inversion. The reason for this choice is that acoustic impedance information exhibits a good agreement with sonic logs at well locations. Also acoustic impedance data has much finer horizontal resolution in comparison with the horizontally sparse log data.

After have been properly conditioned and subjected to the Dix conversion (Dix, 1955), staking velocities were used for the purpose of generating the velocity cube for the first segment. Staking velocities comprised the only velocity data in this segment. By combining these two segments into one velocity model, a single velocity cube is built from the surface (time 0 of seismic) to horizon B (i.e. first and second segments) in our survey.

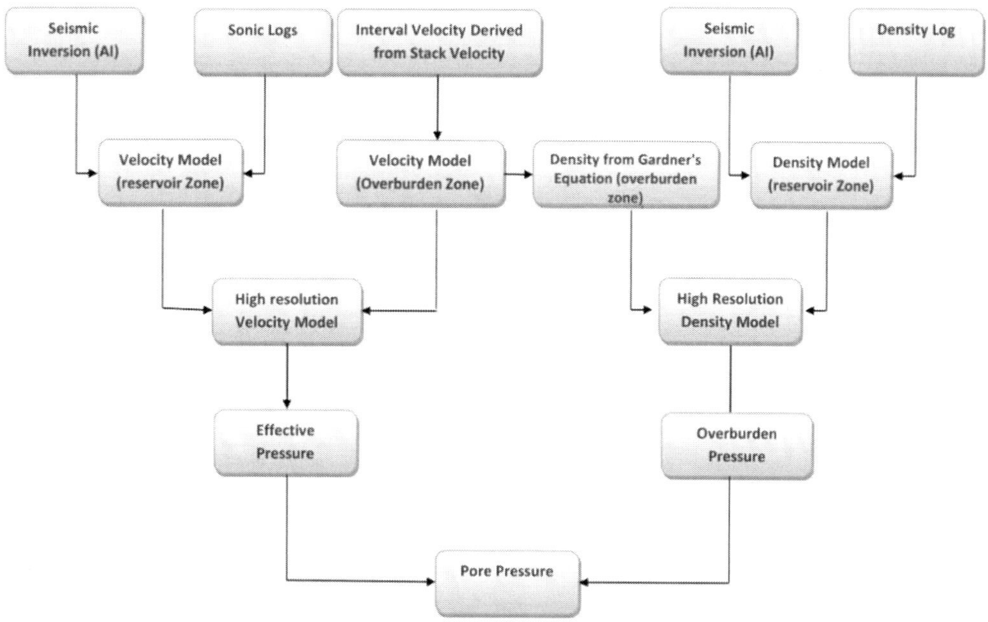

Figure 11. Pore pressure prediction workflow.

Now let's look at the described method in more detail. As the first step the seismic data inverted to obtain acoustic impedances (AI) and subsequently, the obtained results will be used in constructing the velocity and density cube in the reservoir.

Inversion of the seismic data can be performed on both pre-stack and post-stacked data. However, in this study, stacked seismic data yields satisfactory results. As suggested by Khazanehdari and Dutta (2006) spars-spike-type algorithm is used for inverting seismic data. This method employs full stack seismic data, initial geological interpretation, and well log data to generate acoustic impedance.

Moreover, AI data is richer in frequency content than seismic data and hence, it could be a great aid in interpretation and picking the reflectors.

Based on simplified Terzaghi's equation 6, effective pressure is obtained by subtracting overburden pressure from pore pressure. According to Bowers's equation 15, in order to calculate effective pressure, one must have a reliable interval velocity model.

Before using stacking velocities in the first segment, they should be converted to interval velocities using Dix equation (Dix, 1955) which is an equation used to calculate the interval velocity within a series of flat, parallel layers (equation 23).

$$V_{int} = [(t_2 V_{rms2}^2 - t_1 V_{rms1}^2) / (t_2 - t_1)]^{1/2} \tag{23}$$

where V_{int} is interval velocity, t_1, V_{rms1} and t_2 , V_{rms2} are travel time and root mean square velocity of the first and second reflector respectively.

It should be noted that only after performing Dix inversion, the stacking velocities can be used as an approximation to rock velocity. However, due to assumptions such as layered media and hyperbolic moveout, velocities obtained from processing of seismic reflection data often lack the spatial resolution and accuracy.

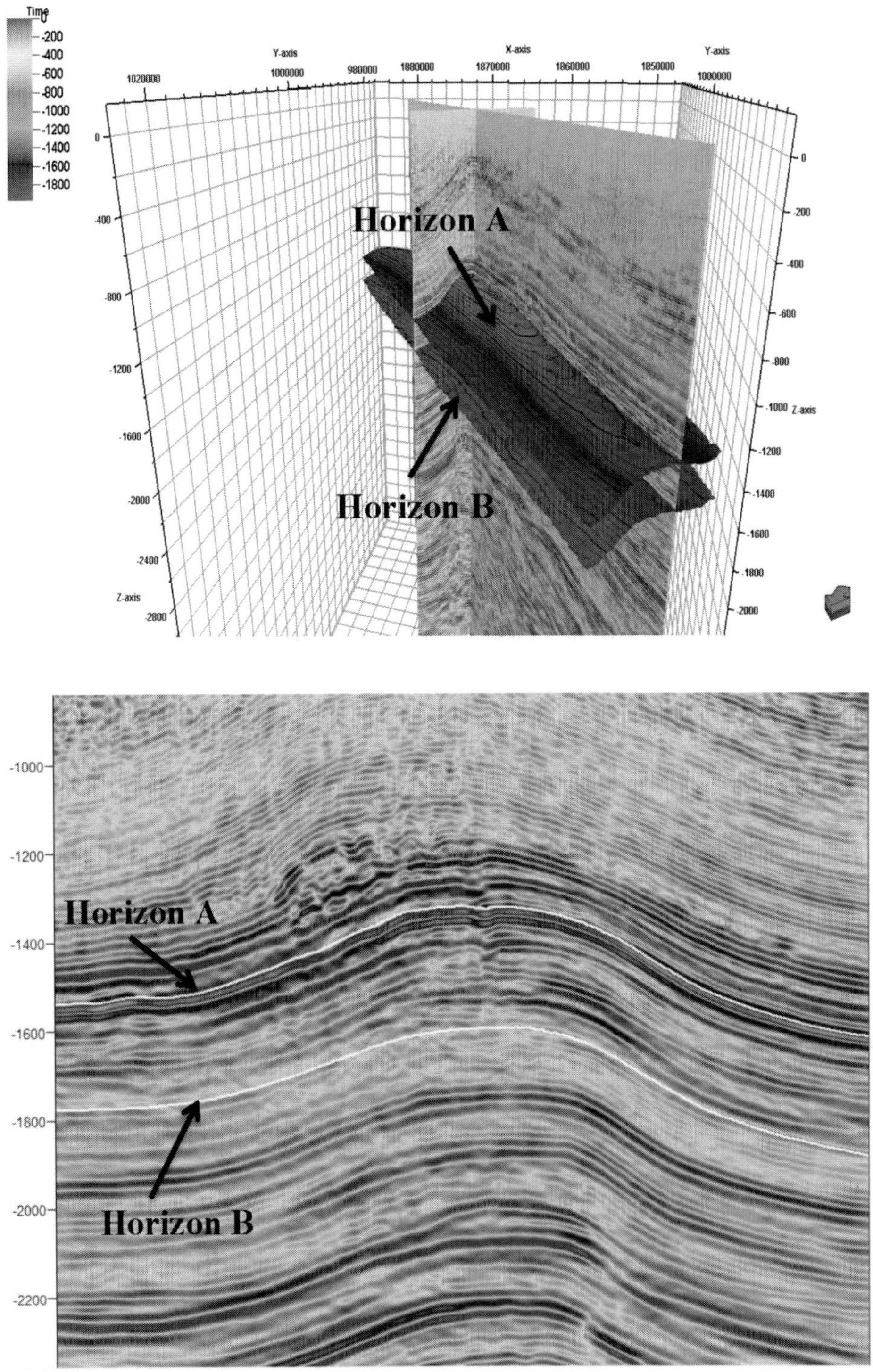

Figure 12. Horizon A and B shown on the seismic data. 3-D view of the horizons is shown in the above Figure, while the 2-D section of the same horizons is depicted in the bottom Figure.

Therefore, it is not always reasonable to assume stack velocity as a valid representative of rock velocity. Figure 13 shows the raw stacking velocities for entire seismic cube (above). Figure in bottom shows the Dix converted and scaled-up stacking velocities in first part of the model.

For the second segment, co-kriging technique is used. Sonic logs of all 35 wells are used as the main data and acoustic impedances from inversion of seismic data are used as the secondary source of data in the co-kriging technique. One of the required information in using geostatistical methods is to obtain valid variograms and detect possible anisotropy in data. Figure 14 shows the variogram map for the sonic log data. This Figure shows that there is anisotropy existent N-W, S-E direction.

The final velocity model is constructed by putting the first and second segments next to each other in their corresponding vertical position as shown in Figure 15. By using this method, higher velocity resolution is obtained from horizon A to horizon B comparing with the generated model which used Dix converted stacking velocities only. This results in higher resolution in final pore pressure estimation. Figure 16 shows the mentioned resolution contrast in the first and second segments by displaying a cross section through the final velocity model and the velocity cube generated from stacking velocities in first section.

To generate the density cube, the model is divided in two similar segments as is the velocity cube. In the first segment, densities from Grardner's equation 14 (which is calibrated by available sonic and density logs in the second segment) are applied.

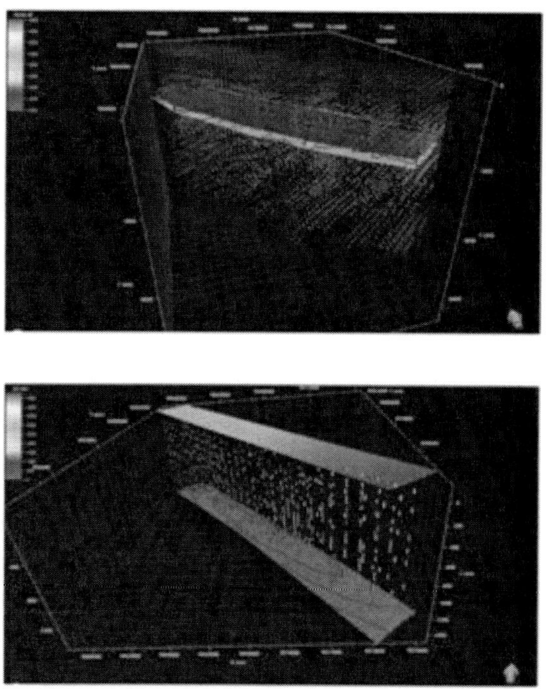

Figure 13. The raw stacking velocities for entire seismic cube (above). Figure at the bottom shows the Dix converted and scaled-up stacking velocities used in the first part of the model.

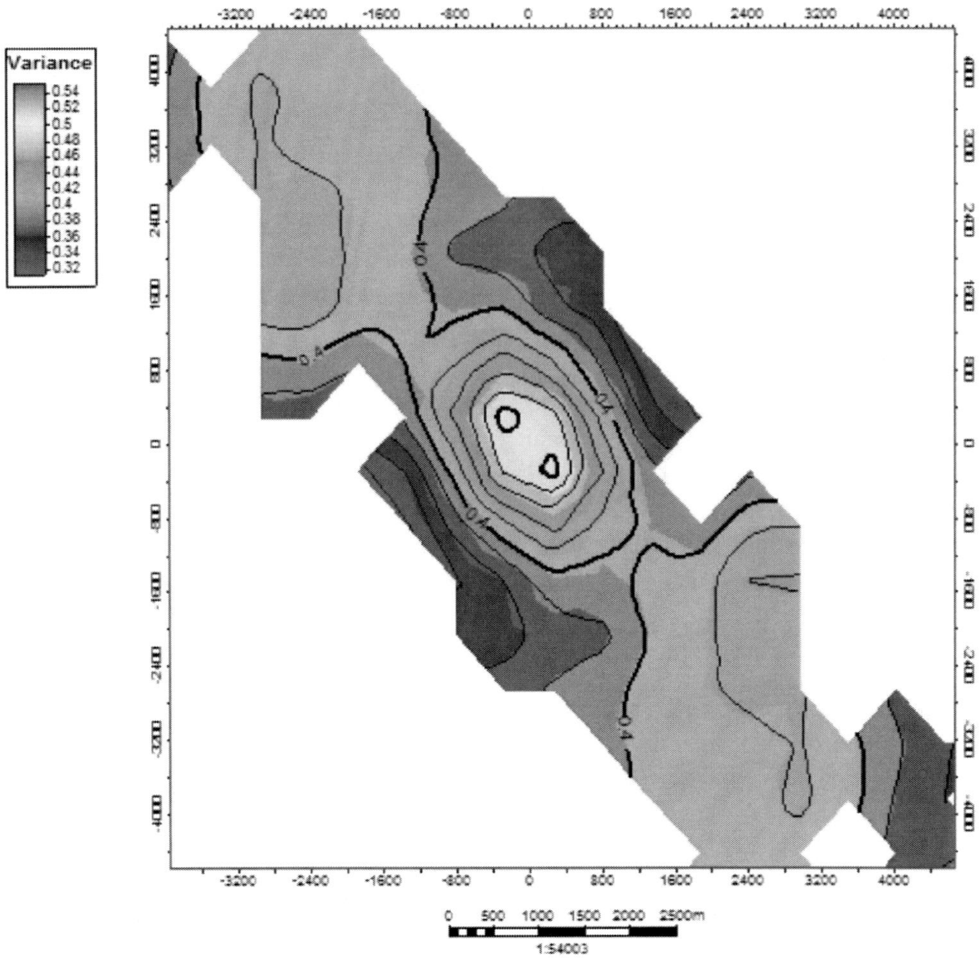

Figure 14. The variogram map Dix for the sonic log data. According to this Figure, anisotropy existents NW-SE direction.

This equation converts seismic interval velocity to density. The calibrated Grardner's equation 14 for this field is:

$$\rho = 0.112V^{0.372} \tag{24}$$

To generate the density cube for the second segment of the model, co-kriging technique is applied to interpolate density logs in offset wells, along with acoustic impedances from inversion of seismic data as the secondary source (Figure 17). Finally, using equation 2, overburden pressure cube produced from the density cube (Figure 17).

The results of the RFT (Repeat Formation Test) in this field were available for six wells. By calculating overburden pressure at locations with known pressure values (RFT results) one can obtain effective pressure at the selected wells. In addition, through extracting seismic velocities at those points, a cross-plot for effective pressure versus velocity in selected well locations can be introduced. This cross-plot can be used to calibrate Bowers's equation (Equation 15).

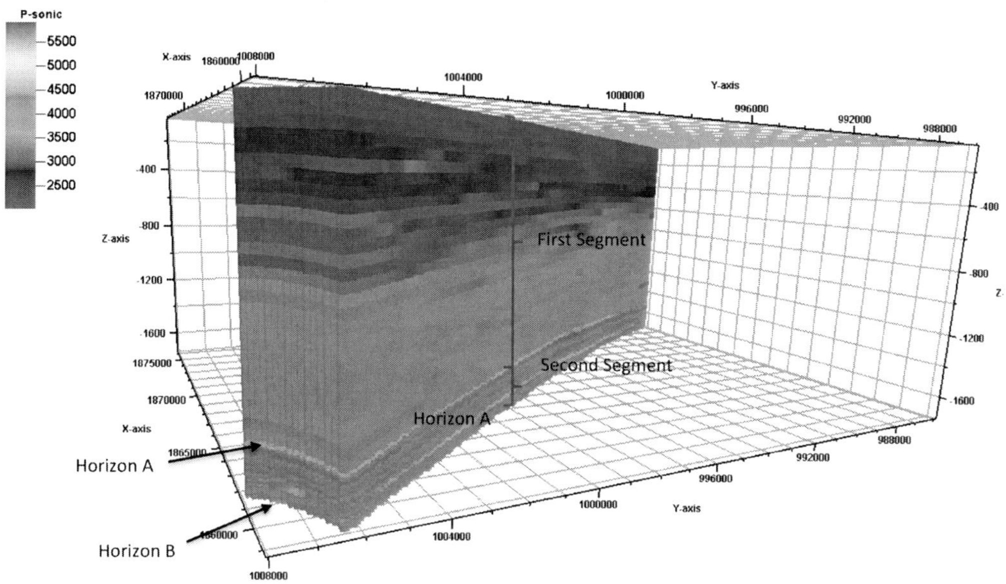

Figure 15. Final generated velocity model. This velocity model constitutes from two segments. The first segment ranges from datum to top of the horizon A and the second segment is the interval between A to horizon B (Reservoir segment). Since the only available data in first segment are the stacking velocities, we use them as our only source of velocity data in this segment. For this reason, stacking velocities converted to interval velocities using Dix equation. Availability of 35 sonic logs in the second segment of the model enabled us to generate the velocity model for the second segment using co-kriging method. Since the result of acoustic inversion (AI) of the seismic data exhibits a good correlation with sonic logs, we used AI data as our secondary source of data in co-kriging technique.

Bowers's equation (Equation 15) is used to create a relationship between velocity data and effective pressure; equation 15 can be rearranged in the following form:

$$V = V_0 + A\sigma_{\text{effective}}^{B} \tag{25}$$

In this equation, V_0 is velocity at zero effective pressure (m/s), V and $\sigma_{\text{effective}}$ are velocity and effective pressure respectively at the desired depth in (psi) and (m/s). A and B are constants of Bowers's equation (Equation 15).

The constant V0 determined as the first step in calibrating Bowers's equation (Equation 15). According to the stacking velocity data, V_0 is 1800 (m/s). After calibration of Bowers's equation (Equation 15), the constants A and B were calculated as 3.678 and 0.743 respectively, using aforementioned procedure.

In the next step, calibrated Bowers's equation (Equation 15) and velocity cube are used to generate effective pressure cube (Figure 18). Then, using simplified Terzaghi's equation (Equation 6) and the obtained effective pressure cube, a pore pressure cube will be produced (Figure 19).

In the final pressure cube, abnormally low pressure region can be observed, which is related to high-velocity rocks (i.e. the cap rock). It should be noted that petrological changes can affect the results of pore pressure prediction. Therefore, these issues should be considered in final pressure map interpretation.

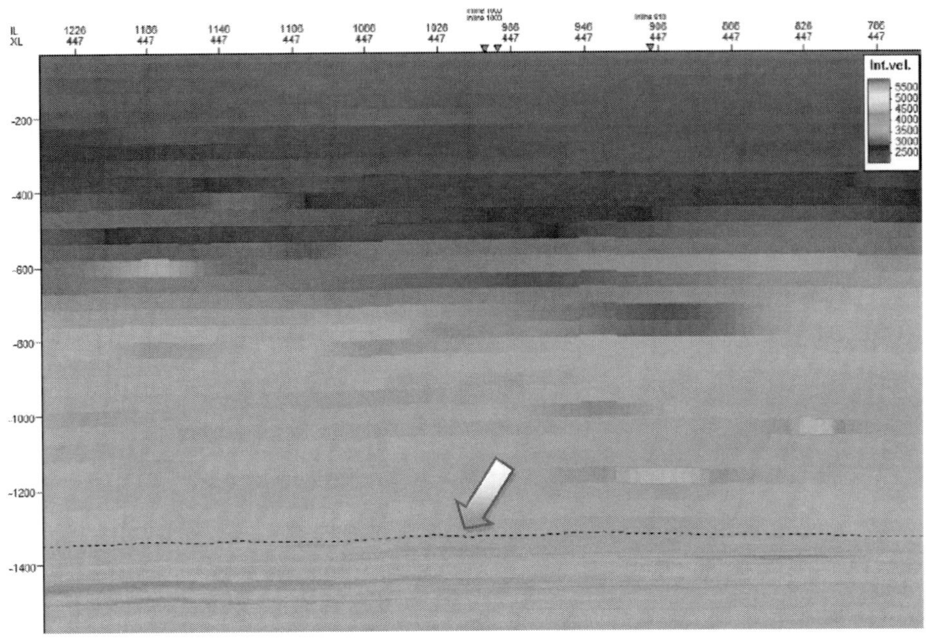

Figure 16. Cross section through velocity cubes obtained from stacking velocities (first section) and high-resolution velocity cube (second section). The results shows that the, interpolation of well logs using geostatistical methods yields a higher resolution in the second segment. Therefore, this velocity cube is more appropriate for pressure prediction. Arrows shows the location of the horizon A which shows the beginning of the second part of the model.

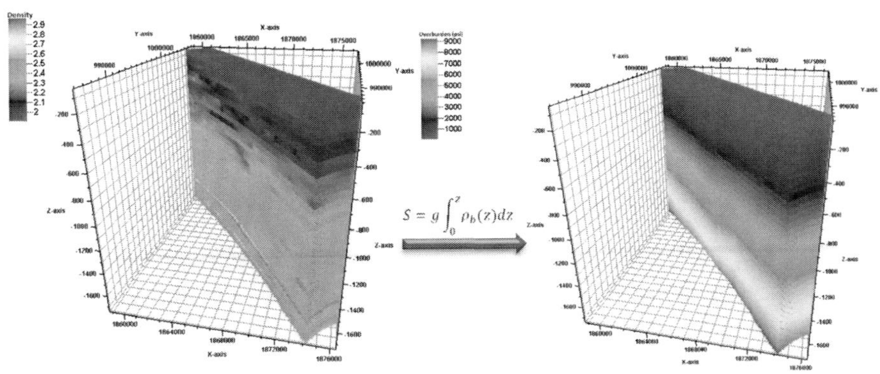

Figure 17. Final density model used in the pore pressure study. Similar to velocity cube in Figure 2, this model we have two different segments with different density sources. In the first segment, interval velocities which obtained from stacking velocities converted to density data using calibrated Grardner's equation 24. Since we had 35 wells with density logs, we interpolate them using geostatistical methods like co-kriging technique. Since AI data shows a good consistence with density data we used it as our secondary source of information in Co-kriging technique. Also AI data have higher horizontal resolution in compare with offset wells. Finally, these two segments set next each other to build a complete density model. Overburden pressure cube shown in the right image. The overburden pressure is the pressure from the weight of rocks and their pore fluid at particular depth. This cube is generated using equation 2 from densities shown in left image.

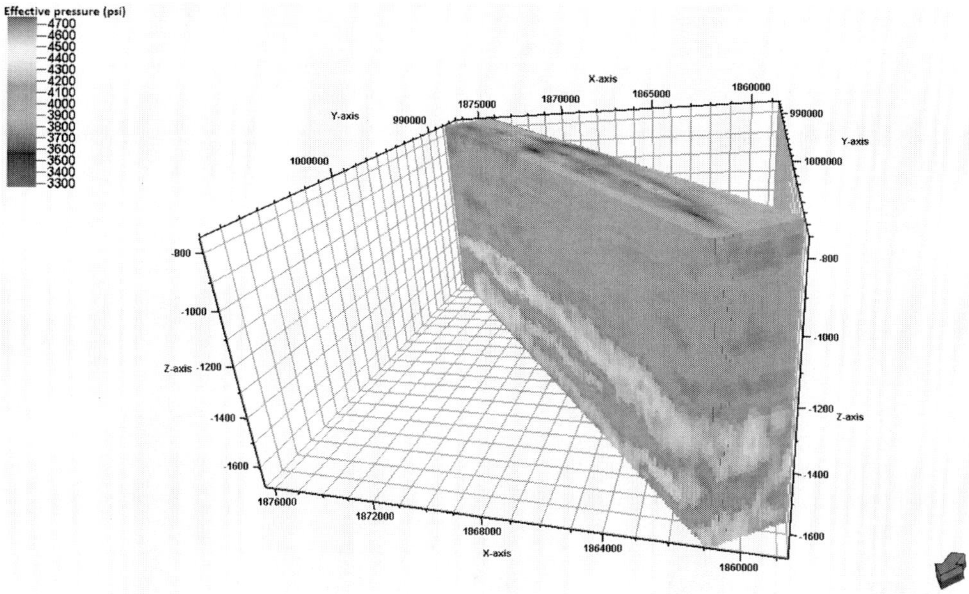

Figure 18. Effective pressure cube at second segment. This cube is generated using calibrated Bowers's equation (Equation 15) from velocity cube (Figure 2). The effective pressure is the pressure acting on solid frame of the rock.

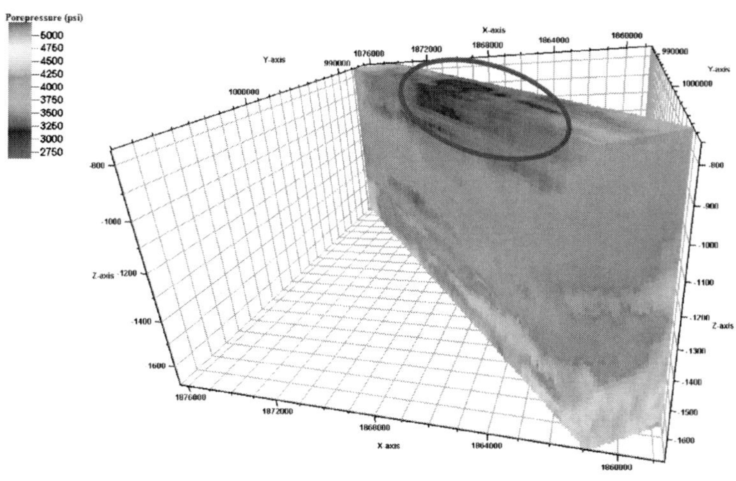

Figure 19. Pore pressure cube from the horizon A to horizon B. This cube is produced using Terzaghi equation (Equation 6). Generally, the estimated pore pressure exhibit a gradual increase with depth and no significant abnormal pressure changes observed in final results. Red ellipse presents a wrong low pressure area in our estimation. This error is results from high velocity zone in cap rock and should be mislaid in interpretation.

To verify the validity of the obtained results, we used blind testing the final results. For this reason, one of the wells containing RFT information was omitted from the workflow for the final validation of the predicted pore pressure (Figure 20). The predicted pressure values (solid line) are in good agreement with RFT data (blue dots).

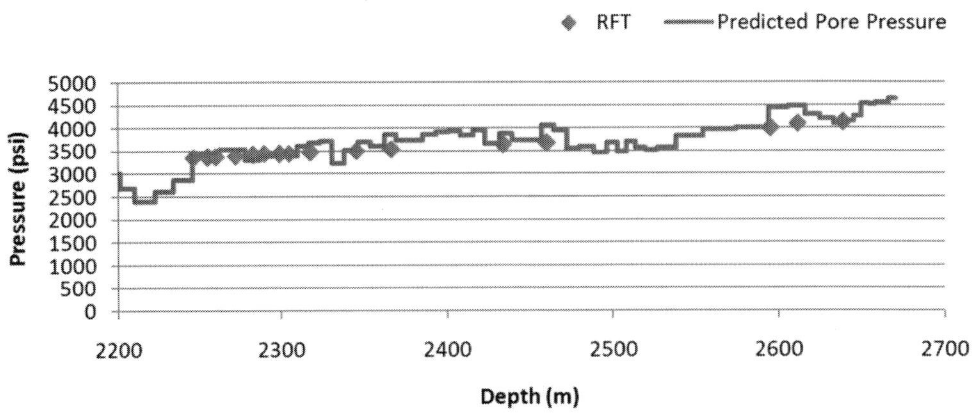

Figure 20. Comparison between pore pressure values obtained from RFT (repetitive formation test) (blue dots) and estimated pore pressure (solid red line). Notice the sudden drop in estimated pressure in below 2250 m, which is the result of the high-velocity zone of cap rock. This pseudo low pressure should be considered as a non-real pressure drop at the seal of the reservoir.

In addition, in Figure 20 the pseudo-drop of pore pressure in depths below 2250 m is evident and reflects the occurred change of the lithology (shale and anhydrite interbedding).

Since accuracy of the derived interval velocities from seismic velocity analysis is greatly affected by seismic quality and the depth of the target zone, it could be a major source of uncertainty in prediction of the pore pressure.

Moreover, it should be mentioned that lateral variation of overburden pressure can also cause major errors in final pressure estimation base on seismic velocities.

Even though these potential pitfalls in using stacking velocities are obstacles in using this information, a good velocity conditioning using VSP or other sources of secondary velocity data can improve the accuracy of the final obtained model.

In the presented workflow, after proper conditioning, seismic derived velocities, which constitute the upper part of the target zone, are used as the only source of velocity data in the first segment of the velocity model.

On the other hand, in the second segment of the model the data from available several sonic logs (from 35 wells) are used in determination of pressure and density parameters. By using sonic and density logs in conjunction with acoustic impedances from inversion of the post stack seismic data, the velocity and density model is generated for the second segment of the model by applying co-kriging technique.

It is shown that the accuracy and resolution of the velocity model could be greatly enhanced by using available well logs and a proper geostatistical method. This high resolution velocity model results in high resolution pore pressure model which reveals a lot of information concerning reservoir integrity and characterization. The effective pressure is calculated using Bowers's method which is calibrated using measured pressure data at some well locations and at the end; the pore pressure is calculated using simplified Terzaghi's equation (Equation 6).

The final pore pressure map reveals that no major pressure anomaly exists within the second segment. In addition, the observed low pressure anomaly in the beginning of the second segment is not due to a real drop in pressure (base on measured pressure in wells).

In fact, this error is introduced due to change in formation petrology (shale and anhydride interbedding) at the reservoir seal. It should be noted that the geological study prior to pressure interpretation is an imperative stage in pressure estimation.

Comparison of the predicted pore pressure with the measured pressure at well location which was not included in calibration phase reveals that final pressure estimation is reliable and it is in good agreement with the measured pressure data.

The high resolution pressure data is suitable for well planning and provides detailed information on pressure variation along the future well trajectories. Also, the 3-D pore pressure cube, effective pressure, overburden pressure and fracture pressures are obtained and used for regional geomechanical and pressure assessments.

REFERENCES

Al-Chalabi, M., (1994). Seismic velocity- A critique. *First Break* 12, 589-596.

Athy, L. F., 1930. Density, porosity and compactation of sedimentary rocks. *AAPG Bulletin* 14, 1-24.

Bowers, G. L., (1995). Pore pressure estimation from velocity data: accounting for overpressure mechanism beside undercompactions. *SPE Drilling and Complication* 10, 89-95.

Bowers, G. L., (2002). Detecting high overpressure. *The Leading Edge* 21, 174–177.

Brown, K. E., (1967). *Gas lift Theory and Practice.* Tulsa, OK: Petroleum Publishing Co 924pp.

Bruce, B. and G. L. Bowers, (2002). Pore pressure terminology. *The Leading Edge* 21, 170-173.

Deming, D., C. Cranganu and Y. Lee., (2002). Self-sealing in sedimentary basins. *Journal of Geophysical Research*, 107, 2329.

Chopra, S. and A. Huffman, (2006). Velocity determination for pore pressure prediction. *The Leading Edge* 25/12, 1502-1515.

Cranganu, C., (2007). Using artificial neural networks to predict the presence of overpressured zones in the anadarko basin, Oklahoma. *Pure and Applied Geophysics* 64, 2067 - 2081

Dix, C. H., (1955). Seismic velocities from surface measurements. *Geophysics* 20, 68-66.

Dutta, N. C. and J. Khazanedari, (2006). Estimation of formation fluid pressure using high-resolution velocity from inversion of seismic data and a rock physics model based on compaction and burial diagenesis of shales. *The Leading Edge* 25, 1528-1539.

Dutta, N. C., (2002). Geopressure prediction using seismic data. Current status and the road ahead. *Geophysics* 67, 2012-2041.

Dutta, N. C., (1987). Geopressure. *Geophysics Reprint Series, No. 7. Society of Exploration Geophysicists* 365pp.

Eaton, B. A., (1972). Graphical method predicts geopressure worldwide. *World Oil* 6, 51-56.

Fertl,W. H., (1976). *Abnormal formation pressure.* Elsevier, Amsterdam 385.

Fertl, W. H., R. E. Chapman and R. F. Holz, (1994). *Studies in abnormal pressure.* Elsevier, Amsterdam 251-2790.

Gardner, L. W. and A. R. Gregory, (1974). Formation velocity and density- The diagnostic basics of stratigraphic traps. *Geophysics* 39, 770–780.

Gregory, A. R., (1977). Aspects of rock physics from laboratory and log data that are important to seismic interpretation. *AAPG memoir* 26, 15-46

Khazanehdari, J. and N. Dutta, (2006). High-resolution pore pressure prediction using seismic inversion and velocity analysis. *SEG/New Orleans 2006 Annual Meeting* 1540-1544.

Magara, K., (1975). Importance of aquathermal pressure effect in gulf coast. *AAPG Bull.* 68, 673-683.

Pennebaker, E. S., (1968). Seismic data depth magnitude of abnormal pressure. *Wold Oil* 166, 73-77.

Pickett, G. R. (1963). Acoustic character logs and their applications information evaluation. *Journal of Petroleum Technology* 15, 650-667.

Sayers, C. M., (2006). An introduction to velocity-based pore-pressure estimation. *The Leading Edge* 25, 1496-1500.

Soleymani, H. and M. A. Riahi, (2012). Velocity Based Pore Pressure Prediction: A Case Study at one of the Iranian South West Oil Fields. *Journal of Petroleum Science and Engineering* 94-95, 40-46.

Stump, B., P. B. Flemings., T. Finkbeiner., and M. D. Zoback, (1998). Pressure differences between overpressure ands and bounding shales of the Eugene Island 300 field (offshore Louisiana, US.) with implications for fluid flow induced by sediment loading: Presented at Overpressures in Petroleum Exploration Conf.

Tanner, M. T. and F. Koehler, (1969). Velocity spectra - digital computer derivation and applications of velocity functions. *Geophysics* 34, 859-881.

Terzaghi, K. and R. B. Peck, (1948). *Soil mechanics in engineering practice.* Wiley, New York, NY 566pp.

Tosaya, C. A., (1982). Acoustical properties of clay bearing rocks. *PhD dissertation,* Stanford University.

The last section of this chapter (Practical Example of Velocity Based Pore Pressure Prediction) is an extended review of the paper by *Soleymani* and *Riahi* (2012). The author of this section massively reused the contents of the above-mentioned paper.

INDEX

C

D

G

H

I

M

N

O

P

Q

R

S

T

U

V

W

X

Y